Y0-DEN-860

CELLULAR MARKETING

Robert Steuernagel

QUANTUM PUBLISHING

Editor and Publisher *Paul Mandelstein*
Project Management, Electronic Composition,
  and Technical Illustrations *Professional Book Center*
Design *Lee Ballentine*
Copyediting *David W. Rich*
Proofreading *Caryl Riedel*

Copyright © 1993 Quantum Publishing, Inc.
All rights reserved. No part of this book may be reproduced in any form, nor incorporated in electronic data-retrieval systems, without written permission from the publisher.

Published by Quantum Publishing, Inc.
P.O. Box 310
Mendocino, California 95460

Library of Congress Cataloging-in-Publication Data
  Steuernagel, Robert
    Cellular marketing / Robert Steuernagel
      p. cm.
    Includes index.
    ISBN 0-930633-20-2 : $79.00
    1. Cellular radio equipment industry—United States—Marketing.
  I. Title
  HD9696.R363U675   1993
  621.3845'6'0688—dc20   93-15319
                                                     CIP

ISBN 0-930633-20-2
Printed in the United States of America

Brand and product names referenced in this book are trademarks or registered trademarks of their respective holders and are used here for informational purposes only. The author and publisher gratefully acknowledge the following entities for permission to reproduce the following figures: Figure 2-2, Motorola Inc.; Figure 2-3, Oki-Telecom; Figure 2-4, Nokia® mobile phones.

CONTENTS

FOREWORD xi

ABOUT THE AUTHOR xv

1 **INTRODUCTION** 1
  The Cellular Marketplace   1
  Cellular Marketing History   3
  The Industry   6
  The Marketing Framework   6
  Restrictions on Bell Companies   7

2 **MARKET AND PRODUCT DEFINITION** 9
  Customer Benefits   9

Market Segmentation   10
    Market Changes   12
    Buyers versus Users   13
Product Attributes   14
Telephones   15
Services versus Products   17
Service Features   19
Other Product Elements   20
Further Market Segmentation   22

3   THE MARKETING ORGANIZATION 25

The Evolution of Sales   25
    Channel Diversification   26
    Increased Role of Marketing   27
The Marketing Function   28
Decentralization   31
Centralization   32

**4   THE STRATEGIC PLAN AND THE
MARKETING PLAN 35**

The Strategic Plan   36
The Corporate Mission and Objectives   37
Sample Statements   38
    The Mission   38
    Strategic Objectives   39
    Strategies   39
Marketing Objectives and Pricing   40
    Strategies to Achieve Marketing Objectives   40
    Business Assumptions for Pricing   41
    Specific Pricing Structure Assumptions   42
    *Complementary Marketing Strategies
     and Assumptions*   43
The Annual Marketing Plan   44

5 PRICING 49

Pricing Development   50

Pricing and the Marketing Mix   51

Pricing Structure and Levels   52
- *Multiple Pricing Plans*   54
- *Comparative Plan Features*   55
- *Pricing Pitfalls*   55
- *Revenue Relationships*   57
- *Wholesale Pricing*   57
- *Crossover Point*   58
- *Pricing Plans and Sales Commissions*   60

Vertical Features and Options   60

Roaming   61

Financial Considerations   62
- *Relationships of Pricing Elements*   62
- *Price Elasticity*   63
- *Price Level Strategies*   64
- *Long-Term Outlook for Pricing*   66

6 SALES CHANNELS 67

Agents   67

Dealers   68

Broker-Agents   69

Resellers   70

Retailers   71

Internal Direct Sales Force/Major Accounts   73

Direct Marketing   74

Company Stores   76

Channel Mix   77
- *The Hybrid Model*   77

Market Share by Channel   78

Sales Channel Characteristics   78

Channel Conflict   81

Sales Commissions   81
    Residuals   82

Resellers   83

Agents versus Resellers   85

National Cellular Programs   85

Agents: The Primary Channel   86
    Agent Organization   86
    Agent Sales and Marketing Activities   86

Customers   88

Pricing and Equipment   89

Service   90

Recent Trends   91

Agent Contracts   92

Agency Terms   93
    Commissions   94

The Ideal Agent Contract   94

Agent Financials   97

7 ADVERTISING AND PROMOTION 103

Definitions   103

Promoting Intangibles   103

Promoting the Cellular Phone   104
    The Paintbrush Analogy   105
    Lessons for Promotions   105

Advertising   106
    Educational Advertising   106
    Promotional Advertising   107
    Measurability   108
    Time-Limited Call to Action   109
    Single-Minded Proposition   109

Media   110

Public Relations   111

Advertising Pitfalls   112
Auto-Market Research   113

8 MARKETING MEASUREMENT 115

The Disciplines of Measurement   115
   Congruents   115
   Basic Rules of Management   118
   Relative Channel Performance   119

Useful Measures   119

General Measures of Marketing Success   120

Measuring Promotion Success   122
   Information Gathering   122

Customer Satisfaction   123
   Detail Requirements   124
   Longitudinal Studies   125

Summary   125

9 COMPETITIVE BEHAVIOR AND ANALYSIS . . . 127

Competitive Differentiation   128
Boutique versus Industry   130
Competitive Analysis   131
Unique Customer Proposition   132
Competing on Your Own Terms   133
Who Is the Competitor?   134
Competitive Backlash   135
Flattery   136
Disparagement   136
Security   137
Overvaluing Competitive Information   137
How Fast Must a System Grow?   138

10 THE SALES PROCESS 141

General Considerations   141
- *Coordination*   142
- *Sales Support*   143
- *Lead Generation*   143
- *Campaigns*   143

Direct Sales   144
- *Sales Planning*   144
- *Sales Presentation*   145

Retail Sales   146
- *Different Customers*   146
- *Tactical Management*   146

11 CELLULAR SALES PRODUCTIVITY 149

Current versus Future Revenue   151
- *Channel Income Statement*   151

Calculations   152

Analysis   154
- *Revenue/Cost Ratio*   154
- *Expensive—But Worth the Cost?*   155
- *Matching Segments to Channels*   156
- *Indications for Action*   157
- *Benefits*   157

12 THE LIMITS TO GROWTH 159

The Symptoms   160

The Problem   160

The Solution   161

Growing the Sales Capacity   162

Other Uses   163
- *"Topping Out"*   163
- *The Limits to Growth*   164

13 SELLING THE CONVERSION TO DIGITAL ... 165

The Need for Digital: Capacity 165
  Making Customers Part of the Problem 166
A Solution 167

14 SWORDS INTO PLOWSHARES: Cooperation and the Evolution of PCS 169

A New Approach 169

Cellular 171

Local Exchange Companies (LECs) 172

Equipment Providers 173

Cable Television Interests 174

Paging Carriers 175

New PCS Carrier Startups 175

Conclusion 176

Index 179

FOREWORD

Any attempt to describe the cellular industry is like trying to describe a bullet as it passes by. While change is the watchword for the entire business environment, its effects are multiplied in an industry that is characterized by rapid growth and constant evolution in both technology and markets.

The cellular telephone industry is one of the most exciting with which to be associated. It is one of the few industries that has three of the major elements required for maximum career satisfaction: first, an expanding industry with the potential for advancement, personal growth, and financial reward; second, a high-tech area with all of the associated challenges, excitement, and promise; and third, a product that fills a genuine, urgent customer need.

The cellular industry is relatively naive in marketing, in spite of its phenomenal success. Perhaps its lack of marketing expertise is a legacy of its owners, mostly telephone companies, known for their lack of marketing prowess, or other communications and private companies whose marketing credentials are likewise not their best asset. The industry's lack of marketing ability has certainly not prevented it from gaining customers, but has caused it to pay an enormous price for them. The conventional, sex-biased belief in the industry seems to be that "real men sell; others market," and that marketing is fluff that sales managers do in their spare time. Perhaps this book can fill a gap in the understanding of marketing techniques that can substantially lower the cost of acquiring and keeping cellular customers.

The purpose of this book is to demonstrate how intelligent marketing contributes to cellular success when combined with good sales techniques. While it is written for marketing and sales managers, it is important for managers at all levels and functions in the cellular industry to appreciate how marketing strategy influences, and is affected by, engineering, finance, operations, and so on. Its emphasis is on marketing strategies rather than sales strategies, recognizing that sales is just one important element of the marketing effort. It assumes that the reader has some knowledge of basic marketing techniques and shows how they can be applied effectively to cellular. A general understanding of cellular service and cellular telephones is also assumed.

Many of the statistics and examples used here will differ from the experience of readers. These statistics, such as usage per customer or cost per minute, are based on a mixture of my experience, industry averages, and actual results in specific markets, and may be outdated before they are printed. They should be treated as examples of benchmarks for marketing decisions, with the latest statistics for the reader's market(s) as the most appropriate operating data. My recommendations for cellular marketing operations are based on what has worked for me and what is currently working for others, combined with some things that might work for all in the future and an exclusion of things that have never worked.

No one has all the answers. If this book merely stimulates readers to study the marketing issues explored here more carefully and arrive at their own conclusions, I will have considered it a success.

ABOUT THE AUTHOR

Robert Steuernagel served as Vice President of Marketing at PacTel Cellular, where he was responsible for sales, marketing, and customer service for all of PacTel's markets, and as Vice President and General Manager of Los Angeles Operations, the largest cellular carrier in the country. Prior to AT&T's divestiture, he was Director of Pricing at its Advanced Mobile Phone Service subsidiary. In that capacity he developed the marketing strategy and pricing strategy for all of the Bell markets. He developed the actual prices for many markets and testified as marketing witness before the state commissions in Illinois and California for the first two markets, Chicago and Los Angeles (tied with New York City). He has also served in various marketing positions at Pacific Bell. Most recently he is consulting in cellular/PCS.

Mr. Steuernagel holds a B.S. from Bucknell University and an M.B.A. from Fordham University.

CHAPTER 1

INTRODUCTION

THE CELLULAR MARKETPLACE

While cellular came of age during a period of economic growth (from 1983-88), its results, measured in customer growth, during the recessionary period that followed have been equally impressive.

The Cellular Telecommunications Industry Association (CTIA), the major industry association, reports that the number of cellular subscribers grew more than 45% in 1992. In terms of number of customers, this represents the highest gain ever recorded. It is truly extraordinary when you consider how much harder it is to maintain growth rates when the base is 8 million instead of 1 million, as it was a few years ago (see Figure 1–1).

YEAR	SYSTEMS	SUBSCRIBERS	REVENUE($M)
1984	32	91,600	178
1985	102	340,213	483
1986	166	681,865	853
1987	312	1,230,855	1,152
1988	517	2,069,441	1,959
1989	584	3,508,944	3,341
1990	751	5,283,055	4,549
1991	1,252	7,557,148	5,709
1992	1,506	11,032,753	7,822

Source: Cellular Telecommunications Industry Association

Figure 1-1 *Cellular growth*

Revenue for the same period was up 38% over the previous year. This indicates that while the revenue gain was healthy, the revenue per subscriber was lower for new customers than for the existing base.

The total domestic cellular subscriber base grew to 11.0 million, and the annual gross service revenue grew to $7.8 billion. Merely dividing the total revenue by the average base (and again by 12) gives an estimate of monthly revenue per customer of $70. The actual monthly bill per customer reported was $68.68, down from $72.74 the previous year.

The revenue per customer has been declining almost since cellular began (records going back to 1987 put the monthly figure at $96.83). It is not a phenomenon of economic recession at all. While it is true that many newer markets have smaller areas and fewer customers with the highest need for cellular, large markets report that long-time customers use cellular as much as they always have. The primary reason for lower revenue per customer is that newer customers belong to market segments that find value in cellular, but don't need to use it as much as earlier adopters. Declining cellular telephone prices allow potential customers to overcome entry barriers and subscribe to the service, even though they have lower usage requirements.

The outlook for cellular growth is just as healthy as its history. Originally projected to average a penetration of 1% to 1.5% in a metropolitan area, over 5 years most carriers have achieved a penetration greater than 3% and project longer-term penetrations of 7% to 10%. Various estimates project growth to over 25 million cellular users (which would reflect at least 10% penetration of the U.S. population) and over $25 billion in annual revenues before the end of the decade (in spite of new services like Personal Communications Services and other wireless voice technologies). This will require new technologies in digital cellular that will expand its capacity, improve its performance, offer additional user benefits and services, and lower its costs.

The declining revenue per customer is just one statistic, like the price of cellular phones, that points to the rapid changes in the cellular marketplace over its short history. One of the challenges for cellular marketing, then, is to find ways to lower marketing costs and expand sales channels to newer market segments which produce lower revenue, while providing superior marketing and sales support to high-potential market segments. Such changes are already occurring in many markets.

CELLULAR MARKETING HISTORY

It is difficult to believe that there was a time when the market for cellular was as uncertain as the market for advanced services like wireless data is today. As late as 1982, it was not known whether carriers would sell at retail or wholesale what price cellular customers would be willing to pay and whether the target market for cellular would be defined by the dimensions of vertical markets (industries) or personal lifestyle demographics like income. Mobile telephones have been around for a long time, and many predicted that cellular would merely displace some of the 20,000 or so existing mobile telephones. These users were very unhappy with the service because of busy channels and the poor quality of reception.

As it has happened, however, increases in dependability, ease of use, and capacity set the stage for the real reasons for cellular's success: a lot of new carriers bent on success, an enormous amount of free publicity, and the resulting aggressive marketing of the service. All the marketing would have been for naught if the service had been poor. But the promise of cellular technology was fulfilled in commercial implementation, and early customers confirmed the value in cellular that was promised in the marketing.

Some of the marketing methods used today have a heritage in the business philosophy and regulatory constraints of the original participants. Before AT&T was broken up, the assumption was that it might be the only carrier, and provide service only at wholesale to local mobile phone dealers. After it was decided that there would be two carriers, restrictions on selling CPE (Customer Premises Equipment, or telephones) in combination with services led the telephone companies to believe that retail sales agents would be required to provide both equipment and service subscription through a single vendor (one-stop shopping). The agent concept has survived and flourished, even after such restrictions were removed.

Another regulatory constraint which helped shape cellular marketing was the reseller concept. Because the wireline (WL)[1] licensees of cellular were able to agree on market ownership early, while the diverse ownership of non-wirelines (NWLs) had much to settle before building systems, the FCC agreed to require the WLs to resell cellular service at wholesale. This allowed the NWL in each market to sell service at retail by

---

[1] The FCC awarded two cellular licenses in each metropolitan or other geographic area: one to a telephone company affiliate (wireline) and one to anybody else (non-wireline). The 90 largest markets were awarded on the merits of the applications, but the telephone companies agreed on who would apply for each market rather than compete for them. The non-wirelines, as local, diverse business interests, could come to no such easy agreements.

reselling the wireline carrier's service, until it could begin commercial service later. This was known as the "headstart" provision. It created an additional sales channel for cellular which all carriers might have ignored forever if left to their own judgement. They still often try to suppress it because of its historical relationship to supporting future competitors.

Because cellular licenses were awarded on a market-by-market basis, there were no carriers or other providers with a large enough regional or national market to consider national advertising, sales, or marketing strategy for cellular, except those that had been developed through consolidation or industry affiliation.

As a result of this environment, cellular is marketed and sold on a local basis. Even large carriers like McCaw and PacTel decentralize their marketing and their management. While a decentralized approach is appropriate in today's business environment, it inhibits the development of broad strategies, devalues marketing skills and resources as a small appendage of the sales effort, and limits the ability to communicate the lessons of both successful and failed efforts. As a result, customer gain has been driven by the brute force of sales resources, more than the thoughtful application of broader marketing disciplines.

The legacy of this heritage is that cellular marketing is relatively unsophisticated. While cellular has achieved spectacular growth and acceptance, it has done so at an enormous marketing and sales cost (see Figure 1–2). Cellular's lack of marketing sophistication derives from local orientation, the low marketing skills of its telephone company and other owners, and the short-term focus of decentralized management.

	AGENT	CARRIER
Sales cost	$250	$350
Loaded marketing cost	$400	$550

Figure 1–2   *Estimated marketing cost per cellular activation*

Chapter 1 INTRODUCTION

THE INDUSTRY

The cellular industry has at its core the two carriers in each market and the independent, external sales channels that complement (or compete with) the carrier's own sales efforts. These external channels include resellers, agents, dealers, and retailers.

Manufacturers of cellular telephones and network equipment are the second major element of the industry and influence the features of the service and the sales and value presentation to the customer.

Other major elements of the industry also influence its marketing: billing services; producers of cellular accessories to improve cellular communication and allow new applications of cellular services like FAX or data; and auxiliary service providers like voice messaging, paging, and local and long-distance telecommunications companies.

THE MARKETING FRAMEWORK

We will approach cellular marketing in the classic manner: We will describe the product features and target markets and show how the various elements of the marketing mix (Product, Price, Promotion, Place [Distribution/Sales]) interact with each other to maximize the penetration and satisfaction of the target market at appropriate cost levels. We will explore the fit of marketing with the overall strategic plan and look at the development of the marketing plan as part of the annual operations plan. We will look at marketing and sales in terms of organization and execution. Finally, we will look at some particular problems of the marketing approach to cellular, some current trends in marketing strategies, and some guesses about the future of cellular as part of the broader wireless environment.

RESTRICTIONS ON BELL COMPANIES

The wirelines (not just Bell) were required to offer service at wholesale, so that a second carrier could market at retail until it had its own system. Wirelines are still trying to get this requirement removed now that all systems are up, indicating their desire to discourage resale.

The Bell companies were required to offer cellular or wire CPE (Customer Provided Equipment, or telephones) through a separate subsidiary from the carrier. This was eliminated around 1986.

The Bell companies were prohibited from providing information services until 1991. There is still a question about whether this may be reversed by Congress. The companies can get around this requirement by having a third party control the content, while they provide the transmission.

The Bell companies are prohibited from providing long distance service. They must allow the customer to choose a primary carrier and pass the charges through to the customer's bill without markup.

The Bell companies are also prohibited from manufacturing telephone equipment. Thus they cannot manufacture, or have manufactured, cellular telephones. They may, however, private label other manufacturers' designs.

CHAPTER 2

MARKET AND PRODUCT DEFINITION

CUSTOMER BENEFITS

The mainstream cellular customer sees the benefits of cellular service in two major business-related categories: the ability to stay in touch for important decisions when away from traditional, landline telephones; and the ability to be more productive by converting time on the road (or other non-productive time) into useful telephone time.

The evolution of cellular phones from primarily car-installed mobiles to primarily portables has occurred in a few short years. Customers who originally pictured the benefits in terms of automobile travel time can now envision a much larger set of places and circumstances in which cellular can provide benefits.

If we look at classical marketing's consumer adoption process, in which potential customers go through stages of awareness, interest, evaluation, trial, and adoption, we can examine how it relates to the marketing communications process. Figure 2–1 shows some of the experiences of cellular customers as they go through the stages of the decision-making process. It is important that potential customers have these experiences to lead them to a purchase decision. The marketing communications that help this happen include advertising, which both educates the consumer and provides a time-limited special offer to act. As cellular has matured, the effectiveness of educational advertising has gone down as awareness of cellular through other means (e.g., use by associates) has increased. It is more important that the potential customer have multiple exposures to different kinds of advertising in a single day or week, which can only occur when multiple channels promoting various offers of equipment, benefits, and service are active in the market.

Note that the adoption stage in Figure 2–1 occurs *after* the customer has subscribed to the service. Subscriber-based services like cellular must be distinguished from products, in that products are sold only once. With subscriber services, the customer is given a chance every month to re-evaluate whether the cost/benefit relationship justifies continued use. Therefore, as with other subscriber services and consumable repeat-purchase products, adoption actually occurs after a few months of use.

MARKET SEGMENTATION

The early market research in cellular uncovered a target market that was characterized as the mobile portion of business managers in certain vertical markets (industry categories). It wasn't until cellular was actually offered commercially that the true driving factors characterizing the target market were identified as a mix of personal lifestyle and business characteristics. Apparently it wasn't as important that the potential customer

STAGE	EXPERIENCES OF POTENTIAL SUBSCRIBER
Awareness	See cellular phones used on TV See advertisements in all media Observe users on the road See cellular antennas on cars
Interest	Listen to cellular advertising and notice phones among other items advertised Many peers using cellular when contacting them Multiple exposures to users and ads in a single day Ask questions and actively seek information about cellular
Evaluation	Interpret benefits with reference to their own job and lifestyle Associates and friends with similar lifestyles adopt cellular Experience specific occasions when they wish they had cellular Inquire about a sales presentation Delight with certain models of cellular phones
Trial	Try it themselves through friend or sales presentation Buy under promotional offer Uncertainty over commitment to cellular "Sticker shock" of receiving first bill, if price sensitive
Adoption	Desire and value to keep Cost to keep Opportunity cost to stop (stranded investment in phone or service contract) Can't do without it Wonder how they got along without it Change jobs/move

Figure 2–1 *The stages of the cellular buying process and the related experiences of the potential cellular subscriber*

worked in a certain geographic area of the country in a certain industry, although these characteristics correlated positively with cellular users.

The actual drivers of the prime target market are neither the industry nor the type of work, but the size of the company and the relative position in the firm, combined with certain other demographic information. Thus, the profile of the highest potential customer for cellular, as verified by carrier experience, is as follows:

Male
Age 28-55
Income over $70,000
Company of less than 20 employees
Senior executive of the company
Owner of a luxury car

Some of these characteristics may be obvious, but the difficulty is to identify which characteristics *cause* someone to be a cellular user, rather than be merely correlated with users. The least obvious characteristics are senior executives of very small companies, which apparently arise because these individuals need to take an active role in company field operations and stay constantly in touch for important decisions.

Market Changes

While these characteristics still hold for the highest potential customer prospects for cellular, the entire target market has broadened, increasing the potential market, but making it more difficult to target individual segments. Over 30% of cellular users are now women, and the best cellular customers are middle management of very large firms—a segment which was virtually non-existent when cellular was in its infancy. The range of age and income has broadened significantly, and the association with a luxury car is no longer a prerequisite for a good cellular customer.

Not only has the declining price of cellular telephones attracted a broader base of business users, but it has put cellular within the reach of consumers as well. The problem with the consumer market is that the cellular telephone is inexpensive enough for average consumers to *buy* (because of promotions that lower the effective price of the phone), but the service is still too expensive for most to *use*.

This phenomenon is the major cause of the reduction in the average monthly revenue per cellular customer from $95–$100 to less than $70. Existing customers did not reduce their usage in the slow economy since 1989. Rather, the core of users who have a high need for cellular is now diluted with customers from market segments with an increasingly marginal need for it.

This dilution is an important aspect of segmentation for strategy development. If newer customers from different segments are found to be price-sensitive, for example, it is important to design price strategies that appeal to these segments without lowering prices to the existing base of customers. The same applies to strategies for other segments and elements of the marketing mix.

Buyers versus Users

It is also important to distinguish the characteristics of users from those of buyers. While the profile of the highest-potential cellular buyer still holds, most of them have long since bought and retained cellular service in major markets. The segment has been saturated, and broader segments are now addressed. Even though 30% of users are women, this represents a weighted average of an increasingly penetrated segment, and the current sales rate for the female segment may be closer to 50%. Thus it is important to understand the makeup of the base, but it may not be a timely indicator of the characteristics of today's prime prospect in a fast-changing market like cellular.

While the highest-potential segment for selling cellular may be the executive of a very small company, the best customer is the middle manager of a large company. Fortune 500 companies that have adopted cellular on an institutional basis have the heaviest users and the longest-lived ones. "Institutional basis" means that the company has adopted the concept of cellular as a productivity tool, and pays for the telephone and the service for executives in the organization according to function and management level rather than individual need. Such users are not price-sensitive (since the company pays the bill), are long-lived because such managers stay with the company longer, and thus constitute a market segment with a much better-than-average usage per month and subscriber life. Such corporate customers are often referred to as "major accounts," based on the way the account buys versus the behavior of the individual corporate user, who may never see a salesperson.

Major accounts usually buy from a carrier just because they assume that the best service will come from the carrier as a principal provider. However, they will often issue an RFP (Request For Proposal) and may get a better price from a reseller. They can't get a better price from an agent, who can only offer the carrier's retail prices. They don't often buy on a national or regional basis (it is rarely offered to them because only McCaw has a coordinated national accounts program). They are interested in service features such as billing and pricing discounts, described later.

PRODUCT ATTRIBUTES

Cellular carriers and resellers would like to believe that the coverage, reliability, customer service, billing, and features of their service are the prime reasons for the adoption of cellular and the selection of the carrier. This, however, is not the case. While some of these elements may influence some individual customers, and promotions of free airtime or other incentives may sometimes affect the choice of carrier, in large measure *the cellular customer focuses the cellular purchase decision on which telephone to buy.*

While the cellular carrier rightly believes that the service decision is more important than the telephone, customer behavior continues to show that the intangibility of the service and the physical presence of the telephone push the decision toward the latter. The customer then subscribes to whichever service the vendor of the telephone promotes. The fact that the customer subordinates the carrier decision to the telephone decision is one of the most important aspects of promotional advertising and sales channel decisions, as we will investigate.

TELEPHONES

Cellular telephones can be divided into three categories: mobiles (car-installed, full-power telephones, Figure 2–2), transportables (full-power, large portables, Figure 2–3), and portables (small or pocket-sized, reduced-power portables, Figure 2–4). Originally priced in 1983-84 at about $2,600, the standard cellular mobile phone had a retail price of about $600 by 1991, and in 1992 could be found with a minimum of effort installed for under $300. In 1985, when mobile prices had been

Figure 2–2 *Installed cellular telephone, or "mobile"*

Figure 2–3 *A "transportable" cellular telephone*

reduced to about $1,800, Motorola introduced the first real portable at about $3,000, and its various improved versions have become such a standard item that it has been nicknamed "the brick" because of its size. By 1992 it could be found at a price of around $400, whereas true pocket-sized portables were under $800.

The penetration of portables has been limited by their lower power output, which made them ineffective in many areas of new cellular systems, as well as their price. However, since most cellular systems have been "filled in" with more cell sites, portables now provide satisfactory performance in most metropolitan areas. Combined with price decreases, the sales rate of portables versus mobiles and transportables went from about 15% in 1988 to 70% or more of all cellular telephone sales in 1992.

The implication of the increased penetration of portables means that the role of technical help in installation and maintenance is reduced, and that cellular phones can be sold as a self-service item in department stores and other retail outlets without trained salespeople or technical support. This has implications for sales channel organization, but it also means that the mechanics of the sale are simpler, and the consumer market will

Figure 2–4 *A true "pocket" portable cellular telephone*

be more strongly attracted. The downside of the consumer market is that the consumer is very price-sensitive to the service, and has very low average usage. The consumer market, at the opposite end of the market spectrum from the corporate user, has the lowest usage and the shortest customer life of all of the market segments in cellular.

SERVICES VERSUS PRODUCTS

In this book the terms product and service are interchangeable when speaking of what is delivered to the customer at a price. But there are very important differences in the characteristics of subscriber services versus manufactured products in how they are marketed and in how the efforts of the marketing process become revenues on the financial statement. As the United States becomes a nation with more of its GNP in services than in manufacturing, it is still difficult to detach the marketing literature from its manufacturing orientation.

In a product-based company, sales of product, rather than service subscription, produce revenue. The sale of product is a direct result of the marketing and sales expense incurred in that period. In a subscriber service like cellular, the expenses associated with a sale are recovered in subscriber revenue over the entire period during which the customer continues to subscribe—the "customer life" of the product. Unlike products, there is no relationship between this month's sales cost and this month's revenue. This month's sales expense produces future revenue over several years (see also Cellular Sales Productivity in a later chapter).

Therefore, a cellular carrier's purposeful reduction in sales expenses, as a short-term measure to increase net income, will lower gross sales for the month, but will appear to have little immediate effect on revenue, perhaps leading to further reductions in sales budgets. However, these measures actually cause a long-term slowdown of revenue growth, which is much more difficult to reverse once it has been allowed to occur.

Another important difference between the marketing of subscriber services versus products is that products are sold just once for the full price agreed. Subscriber services, on the other hand, provide the customer the ability every month to consider the value received for the price paid, and to make the purchase decision or cancel service. Therefore, cellular service providers get the chance every month to either satisfy customers or face losing them, depending on the company's attention to system performance and dedication to customer service.

SERVICE FEATURES

Cellular service, as an intangible, has most of its features built into its pricing schedule. In most cases, the pricing is separated into service initiation, monthly access, airtime usage, and long-distance tolls. Some of the features of cellular that distinguish it from other telecommunications services have been designed to appeal to people in motion, as wireless users are often characterized.

1. Local toll and message charges: Such charges are usually included in the airtime charge. This distinguishes cellular from regular, landline telephone service and is an important feature of cellular. It avoids the uncertainty of what toll charges might be involved in a particular call and permits users to estimate their monthly bill based solely on time used. It simplifies the bill by avoiding toll detail, but, most importantly, it positions cellular users' mobile status as "masters of the metropolitan area" rather than being associated with a fixed location. It recognizes their mobility.

2. Airtime: The cellular user always pays for airtime, whether placing a call or receiving one. While this can be positioned as the price cellular users must pay for being available, it is really an expedient that recognizes that local telephone companies do not want to bill landline customers for airtime when calling cellular users. They

have no facility to do so and don't care to have one. While some experiments with "calling-party-pays" have been successful, it will take time to recognize that cellular service is no longer a premium service for the privileged and should have options to bill the calling party.

Under the current limitations, cellular users are hesitant to give out their telephone number except to select parties and valued contacts. Incoming traffic and call completions to cellular phones are severely limited. The possibility of cellular telephone directories will remain unfeasible until calling-party-pays is widely accepted.

OTHER PRODUCT ELEMENTS

Additional elements of the product dimension of cellular service include the adjunct services, which increase the utility of cellular service and tie the user to the service and the carrier, and services like voice messaging, data, telemetry, fixed cellular service, and FAX, each of which may appeal to completely different customers.

Custom calling features are not as useful to cellular users as they are to landline services, because most cellular calls (about 80%) are outgoing. The cellular phone is off or unattended much of the time, so some form of answering or screening is a necessity for those users who do receive incoming calls. Because cellular users usually have to pay for the airtime on incoming calls, they give their cellular number to few people, treating it as an unlisted number, and use their office phone, secretary, voice messaging, or pager to screen incoming calls. Therefore, one very useful custom calling feature is call forwarding, which allows users to redirect incoming calls to these other services.

Such features are often bundled with premium pricing plans, because they provide little revenue and cannot be sold separately economically. When they are separated, it is usually successful only with a carefully planned telemarketing program. It is embarrassing that many carriers still carry speed dialing as a feature just because it is included in the system software, even though virtually every cellular phone has memory dialing. The last time somebody bought this feature was in 1985.

Voice messaging is a feature that ties users to the service and performs an important adjunct service. The primary use is for telephone answering for users who receive telephone calls. Voice mail, rather than live conversation, is used for the exchange of messages among common subscribers. It is not a primary application as in the office environment, because cellular users do not have a "community of interest," a joint need to communicate among themselves, as do managers of a single company. Because of the benefits of voice messaging (increasing usage, binding users to cellular for productivity, and providing $5-10 per month in additional revenue per user), most cellular carriers have adopted this service.

While some carriers claim very high penetration of voice messaging, a 30-40% penetration results only when the service is bundled with premium access plans or promoted as a 60- to 90-day free trial to new customers. Because of the limited amount of incoming call traffic and the number of people with existing voice messaging services at their home or business, it is difficult to raise penetration above 12-15%. This is still an appreciable penetration, and it is an important service to keep heavy users' loyalty.

Billing is often considered an accounting or operations function, but it also is an integral part of the product. A bill that is easy to understand is a benefit to the individual user. The provision of custom billing services to multi-user corporate customers by permitting easy accounting by department or by providing the bill in tape or floppy-disk format for analysis is a competitive advantage in keeping the account.

Additional dimensions of the product definition include the development of products like cellular data transmission and other cellular-based services for special markets. While these markets are just emerging, they will play an increasing role in cellular as it attempts to penetrate additional high-payoff markets.

FURTHER MARKET SEGMENTATION

A subdivision of the market segmentation for cellular (Figure 2–5) can be termed the early adopters. As we look at the history of cellular, the initial primary market segment was the set of executives of small businesses and professionals (lawyers, doctors, etc.). The next segment to adopt cellular was business managers of larger organizations, especially in the target industries (real estate, construction, and financial services). Third, large corporations adopted cellular on a policy basis, that is, companies adopted cellular and paid for the service and phone for classes of managers, either by level of a management or functional area which requires extensive field work. Finally, the consumer segment has appeared as an important segment.

As each new target market accepts cellular, there are additional dimensions to the segmentation of cellular users based

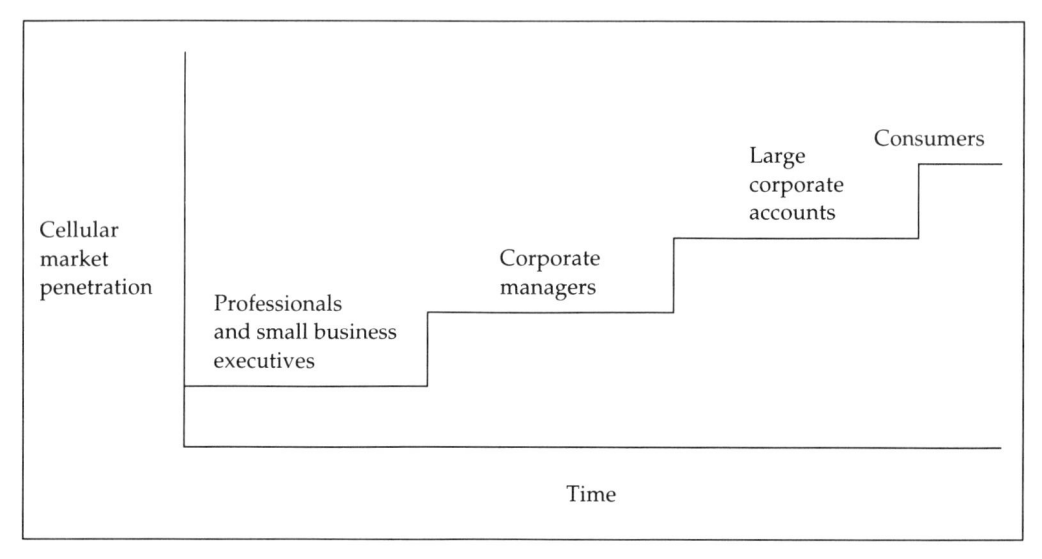

Figure 2–5 *Cellular market segment penetration*

on how they make the decision to employ the service. Thus, within each segment, there are the early adopters. These are people who are comfortable with technological change and are anxious to have the latest electronic devices. The nurturing of the early adopters is important, because these users provide word-of-mouth credibility to new technologies and informally determine the real value as either a fad or a long-term benefit and solution to a problem.

The early adopters of cellular found that the technology was sound, that the cellular telephones were quality units, and that there was true value in productivity gains and communications availability. More important than even these considerations, cellular avoided the dimension in which most new products fail: it was easy to use. It did not require education and user training, and the concept was easy to grasp. It provided users with the opportunity to do things better without asking them to adapt to it. Early adopters in the consumer segment have not always found the same value as business customers. They use cellular less and are more sensitive to price than were early adopters in business.

These characteristics of the developing cellular market lead to an important understanding: Newer cellular systems still must go through the progressive stages of developing users and the confirmation of the service value among early adopters. Thus, newer cellular markets may not immediately jump into an environment of addressing a consumer market through retail channels with the latest in small portable telephones. Each market must go through the stages of technical and marketing maturity, although in newer markets it may occur more rapidly.

These stages include the technical development of the system, as well as marketing development. Portables, for example, do not operate well inside buildings and in fringe areas in newer markets. Initial users of a system must pay careful attention to the installation of a mobile, especially the orientation and type of antenna used on their automobile. As the cellular system builds a user base and generates more usage, more cell sites are required, spaced at closer intervals. These additional cell sites, together with careful power adjustment, channel coordination, and antenna propagation, begin to provide a fairly uniform signal strength, and transportables with lower power and less-carefully oriented antennas provide adequate performance.

Finally, with full system deployment, users find adequate performance with low-power portables inside buildings and in fringe areas, at any orientation of the antenna. Early users of portables, for example, often used them in an awkward posture in order to maintain the antenna in a stationary, vertical position.

CHAPTER 3

THE MARKETING ORGANIZATION

THE EVOLUTION OF SALES

The argument concerning the overall control of revenue production between marketing and sales organizations has never been settled. In most organizations that use a direct sales force, Sales controls revenue and Marketing is a support arm.

For many consumer products, however, Marketing controls revenue, often merely by default because there is no company sales channel. Independent retail stores, the major sales channel members, are not under the control of the companies that produce the products they sell. In these organizations, the product manager may be responsible for the entire income statement for the product division, as well as revenue.

Direct sales forces, or any sales mechanisms in which the manufacturer or service provider controls the sales channels, usually exist where the company (1) has several or several hundred products going to a narrowly defined market; (2) has a product or product line which has a high price (a "big-ticket" item) and/or technical orientation deserving the cost of a direct sales force or other dedicated channel, such as a national chain of specialty stores; (3) has an industrial customer base that prefers to do business with a sales representative in their own office; or (4) some mixture of all of these.

Cellular interests recognized early on that core customers would be mostly business people, who would require a professional salesperson to visit in order to understand the benefits of cellular service. The sale would require not only cellular subscription, but the purchase of a $2,600 cellular phone and technical expertise in installing it correctly and with care respecting the high-end vehicles in which it would be a permanent fixture.

Thus, cellular has had all the elements indicating a need for a direct sales force. Some carriers developed an internal sales force and installation capability, while most found benefit in the participation of contracted sales agents and dealers who would combine sales efforts of cellular service and cellular telephones with professional installation, technical support, and customer service in a single, local, one-stop shopping facility.

As cellular has matured, many of these requirements of sales channels have been modified by changes in technology, customers, and other influences. With cellular phones declining in average retail price from $2,600 to under $600, it became difficult to justify a direct sales force paid high commissions in return for telephone profits. Not only have prices come down, but competition among equipment manufacturers combined with competition for customers have driven the retail profit out of the cellular phone.

Channel Diversification

Increased customer awareness makes it less necessary to convince prospective customers of the benefits of cellular. Many prospects in the consumer and small-business segments are more comfortable buying through retail channels than from a direct sales force, and self-select as prospects rather than require qualification as leads for a salesperson.

Finally, the increased penetration of low-power pocket portables, the wide availability and acceptance of transportables, or full-power phones that can be used without installation, and the availability of quality car installations, all have reduced the importance of installation as an issue. Further improvements in the way cellular phones are programmed, tested, and maintained, plus improvements in their operation because of improved cellular systems and deployment, have diluted the importance of technical support.

Thus, cellular no longer requires a direct sales force to sell to all of its market segments. It is sold through many channels in which it is not the only product or service provided, such as automobile dealerships and consumer electronics stores. Like other products that are sold through multiple channels, the Sales organization is no longer in control of many sales, and Marketing has an increased role in driving customers to channel outlets, rather than Sales driving salespeople to qualified prospects.

Any marketing organization that is controlled by the sales responsibility of a direct sales force will be driven by Sales, and Marketing will usually end up as part of the sales organization. Unfortunately, many sales organizations downplay the value of marketing. Often the marketing function will be a subordinate function of Sales and will be understaffed and little understood and given little autonomous responsibility. With that the Sales organizations's classic portrait of Marketing as a do-nothing staff organization becomes a self-fulfilling prophecy.

Increased Role of Marketing

As the diversity of channels and market segments increases in cellular, Marketing will have a larger role in presenting the service to potential subscribers, and Sales will play a diminished role outside of direct sales. At some point, the ability to understand market segments, the need to match channels to segments, and the need to use "advertising-pull" instead of "sales-push" as the predominant means of putting the prospective customer into a purchase mode take over and make Marketing the dominant, rather than the subordinate, function in the overall marketing/sales cycle.

Advertising-pull refers to using an advertising offer to bring customers who are far along in the marketing awareness stages to a purchase decision or sales situation. Sales-push refers to the need for a direct salesperson to find prospects, nurture leads, and work to bring the customer to a purchase decision through the sales presentation.

In most cellular carrier organizations with some experience in the marketplace, the management of the direct sales function needs to be separate from the management of the broader external sales channels, and the overall marketing function should be separate from both. It should be on equal footing with the overall sales function, if not supervisory to it. The skill set of marketing is entirely different from sales. Marketing must not be a part-time pursuit of the Sales organization.

THE MARKETING FUNCTION

The marketing function is divided into several areas under the general headings of Product Management and Market Management. The mix of functions under these headings, and the subordination of one to the other, is an open issue.

It can be generalized that the function of Market Management is to be the surrogate representative of the customer, embracing market research, market segmentation, and marketing communications: advertising and promotion. The Product Management function is associated with characteristics of the product that satisfy the needs of the market: the design and content of the service, its pricing, and its terms.

In the context of the four Ps of marketing, Product Management comprises the Product and Pricing elements, while

Market Management controls the Promotional element in combination with the understanding of the marketplace in terms of market segmentation and market research. Distribution, the fourth P, is separated into physical delivery, in the case of a manufactured product, and Sales. In cellular, the physical distribution function is limited to the role of customer service in delivering and supporting cellular as an intangible service. Organizationally, Sales and Customer Service usually report separately to a general manager or Marketing and Sales director, because of the size and span of control these functions require.

Where there are only one or few products going to many markets, Market Management dominates the organization and usually controls its direction, and sales are determined by how well diverse markets are managed. Where there are many products and few key market segments, Product Management dominates, and sales are determined by how well each product fits the needs of one or a few key customer segments.

Cellular, with few (or one) products going to few markets, can be successfully organized either way, and will usually be dominated by the relative contribution of each organization (Product Management or Market Management) to sales. Product Management may dominate because of the technical superiority of the cellular system over that of the competitor or good pricing strategy, Market Management may dominate because of its success in advertising and promotion, or its expertise in defining the potential of various market segments. In the best cellular organizations, the two cooperate rather than compete for organizational superiority. Strong personnel in either capacity can also be a deciding factor in taking the lead. Early dominance of the Product Management function in designing all aspects of the cellular offering may, in the maturity of the service, yield to the Market Management function as sub-segmentation, new market penetration, and promotional design gain increasing importance.

Of course, since most companies organize cellular by geographic market, the Sales and Marketing organizations may

30   Chapter 3 THE MARKETING ORGANIZATION

```
                          Sales
                            |
          ┌─────────────────┼─────────────┬──────────┐
       Indirect                         Direct    Customer
          |                                        Service
     ┌────┴────┐
   Retail   Wholesale
```

```
                       Marketing
                           |
           ┌───────────────┴───────────────┐
         Product                         Market
       Management                      Management
           |                               |
      ┌────┴────┐                    ┌─────┴─────┐
   Product    Pricing              Market      Marketing
 Development                      Research  Communications
```

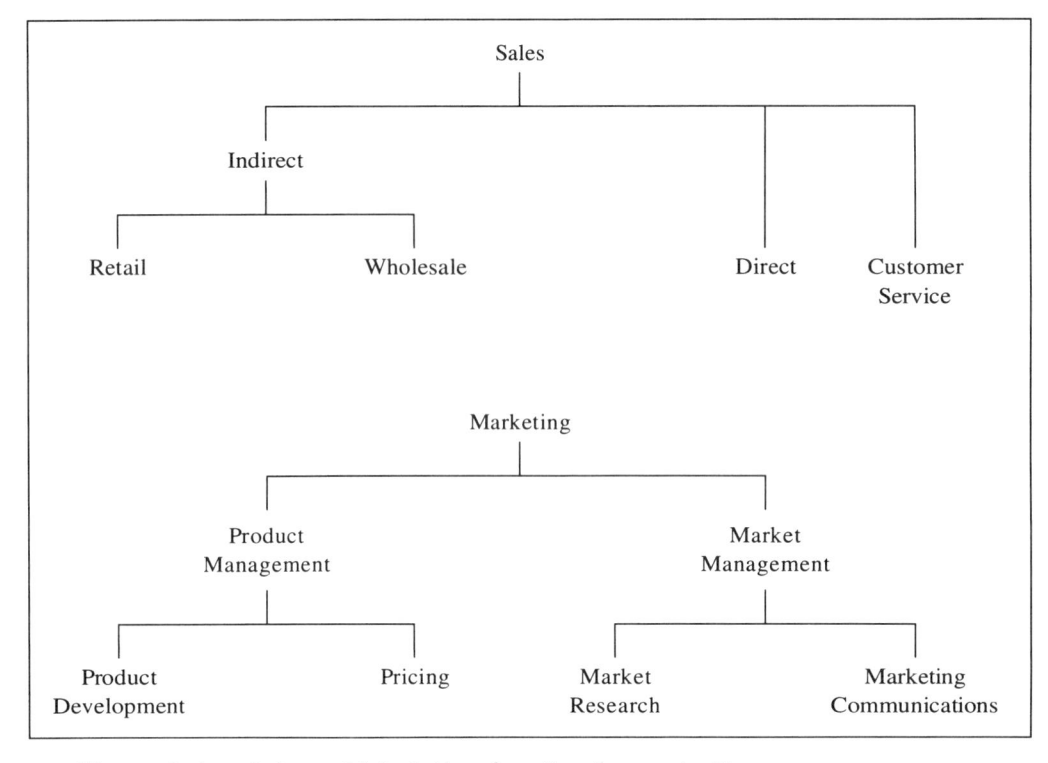

Figure 3–1   *Sales and Marketing functional organization*

begin with only one manager. The organization, no matter how small, must understand the various functions as they relate to the marketing process, and respect the independence and objectivity required when considering the needs and markets of retail versus wholesale channels in the Sales organization, or market research versus product development in the Marketing organization. The local General Manager should at least be able to separate the sales and marketing functions in two different managers. In larger markets, Sales becomes just one element of the overall marketing function under the direction of the senior Marketing executive.

Newer markets are primarily driven by the Sales function. As markets grow and demand a more sophisticated marketing approach, Sales becomes a key subordinate of the overall Marketing organization and function. Some organizations, however, never reach an understanding of the importance of the larger view of Marketing, and retain it as a subordinate and peripheral function of Sales, often viewing it as providing only marginal contribution. This view itself can be a primary cause of a marketing organization's poor performance.

DECENTRALIZATION

Various cellular carriers that have operations in more than a few markets may be organized on a centralized or decentralized basis. Decentralized markets are run based on a local General Manager who has control of most resources and functions for the marketplace. Where functions like human resources or billing are centralized, the General Manager has an influence on the central administration of the function and assumes an allocated portion of the expense.

There are two major advantages with this type of organization: speed and responsibility. The General Manager (GM) has complete P&L (Profit and Loss Statement) accountability for the financial performance of his or her unit, and should have the freedom of responsibility to make the decisions and take action to control it in most areas. Second, the decentralized organization can assess a local problem or opportunity and take immediate steps to correct it or take advantage of it.

Often, the prime areas of concentration for any market are marketing and sales. Accounting, operations, and engineering provide great opportunities for service and efficiency improvements but are fairly straightforward within the capital budget and expense constraints. Marketing and sales provide the greatest variability in terms of opportunities and constraints over a wide range of variables and competitive conditions, requiring creativity and skill. Marketing and Sales managers are usually recruited for General Managers, and the job is about 70% marketing. Because of the P&L responsibilities, financial skills are also a prerequisite for the position, enabling the GM to glean opportunities and corrective actions from the analysis of numeric operating performance data.

The prime disadvantage of a decentralized organization is the lack of depth in any particular skill. Advertising, market planning, public relations, promotion development, pricing, and so forth, are all very shallow, and are usually left to the sales organization as a sideline. These skills become more important as initial core market segments are penetrated, and the need for careful market planning and competitive marketing programs parallels the increasing scarcity of pent-up demand, core market opportunities, ready buyers, and low marketing costs.

CENTRALIZATION

In a centralized organization, each function (marketing/sales, finance, operations, engineering) in a local market reports to a superior at headquarters, and the P&L is consolidated to the company level. Each function has a budget and multi-market performance objectives.

Such an organization has depth of expertise in each functional area, enabling it to create and make use of marketing intelligence with programs that do an excellent job of developing markets and providing superior marketing communications materials, sales training, and so on. But any advantages in such areas are severely constrained by disadvantages. Without any local orientation, marketing programs tailored to a specific market are never available when they are needed, and when available, they are not sensitive to specific, local market needs. Productivity is lost in constantly trying to devote the resources of 10 marketing experts to 10 markets 10% of the time. Most painful is that when one or two of the required resources is available to complete a marketing program, none of the others are. When added to the fact that none of the marketing people are accountable to a local market organization, the problem goes from unwieldy to impossible.

CENTRALIZATION

While no single solution to the problem is available, it appears that the most effective solution is to have a decentralized organization with marketing depth in the headquarters organization. The central organization serves in the following ways:

1. Develops market intelligence and conducts market research specific to each market and consistent among all of them.
2. Assists local markets in developing local expertise and regular promotional programs.
3. Creates promotions and sales support materials of professional quality which are easily adaptable to local markets.
4. Devises and/or supports local marketing programs as pilot programs and new market development efforts before considering broader deployment.
5. Takes a successful, home-grown marketing program from a local market and packages it for additional markets.
6. Investigates new products, applications, and markets, like cellular data transmission, which local markets cannot do effectively.

In such an organization, annual marketing plans are developed by local markets based on assumptions supported by central headquarters' (HQ's) marketing staff. Local markets require bottoms-up market-plan development in order to set their own priorities, develop plans in the context of their overall projected financial performance, and commit to delivering it. The pitfalls of such an organization are in the following categories of problems:

1. The HQ marketing organization is used as an auditor/reviewer of local marketing programs.
2. Local markets resist marketing programs that have been successful in other markets in favor of home-grown pro-

grams that are less sophisticated but "bought into" by local Marketing and Sales managers.

3. The HQ marketing function produces program recommendations that are neither responsive to individual markets nor ready at the time individual markets need them. Where they are available, local markets have no capability to convert them into specific programs.

CHAPTER 4

THE STRATEGIC PLAN AND THE MARKETING PLAN

To managers in many industries and disciplines, planning is merely an activity for idle hands. Poor planning, with poor results and little follow-through, produces poor attitudes about the planning function and a vicious cycle of even less attention to the process. Some argue that plans always change drastically within weeks of their drafting, so why even bother?

For marketing programs to be consistent and rational, in addition to having all the elements of the four Ps working in concert, the programs must be part of larger growth and financial goals and must be part of a larger corporate strategy encompassing all disciplines in the corporation. As important as consistent marketing plans are as a product of a good corporate planning process, marketing also produces the primary quantitative input to the corporate plan: the revenue forecast based on market sizing.

Understanding that plans are made to be changed, the following are several important reasons why strategic plan development is an important part of cellular marketing:

1. A well-constructed plan provides the basis for corporate and management performance measurement.
2. An annual plan provides a benchmark for results and guidelines for corrective action, based on how far actual results deviate from the plan, and whether deviations are caused by market and cost variations or incorrect strategies.
3. The development of a strategic plan is worth at least as much in its process as the resulting document. No other process brings together corporate officers and managers to communicate on basic assumptions and reach consensus on joint focus as much as a team effort to develop the plan.

We will endeavor to outline the formation of a corporate mission, a strategic plan, and the marketing portion of an annual operating plan, in the context of cellular, using hypothetical examples. We will draw on these to show how top-down strategies fit with bottom-up intelligence to provide a coherent, integrated plan for cellular marketing.

THE STRATEGIC PLAN

Ideally, the development of actual marketing and sales programs, advertising strategies, pricing structures, levels, and terms are part of a much broader strategic process. First, the corporate mission, strategic plan and objectives, and marketing plan must be laid out and agreed to by top management. These form the basis of the marketing mix strategy and the pricing plan. The quality of the strategies at each succeeding lower detail level will be a function of the diligence applied to the higher-level strategies on which they are based. The pricing strategy and market forecast are closely intertwined with marketing's quantitative goals (revenue) and the financial goals—not income, cash flow, or other profit measures—and are based on these important marketing inputs. The lower-level marketing and sales strategies and tactics confirm the company's ability to reach these goals. Therefore, the overall financial objectives, along with the marketing strategy, are key to developing a strategic view of each of the marketing functions. The process is iterative. The development of the final marketing plans may require the adjustment of the revenue forecast at the root of the plan. Modified forecasts may change strategies in each area of marketing and their relationship to each other.

THE CORPORATE MISSION AND OBJECTIVES

A cellular carrier might first state its corporate mission as "to become the leading provider of cellular service in the region," and state its objectives in terms of growth and profitability. However, such platitudes provide neither direction to employees nor the subordinate strategies that flow from it. "Leading" does not tell us anything. Leadership must be defined in terms of size, profits, innovative products and promotions, adoption of new technologies, or positions on influential industry forums.

Growth, for example, must be defined in terms of its size, character, and timing. A specific growth goal might be to achieve a level of 100,000 minutes of usage per day within three years. This would avoid the problem of concentrating on customer growth at any cost, perhaps attracting customers with low usage characteristics at high sales costs.

The most crucial test of a strategy or objective is to determine if any different behavior is implied when the strategy is stated in the negative. Thus, since there are no different behaviors implied by *not* becoming the leading cellular carrier, a more specific definition of the goal is required.

It is important to distinguish the difference between strategies and goals. In our terminology, goals are quantitative results to achieve, and strategies are ways to get there. These two terms have been confused even in textbooks. If a goal entails reaching a specific customer count by a certain time, it should include such factors as the market segments to which they belong, the usage they should generate, and the mix of peak and off-peak traffic. These will help determine specific strategies that will get the company there.

A goal to reach a certain penetration of the corporate market segment might define a strategy that includes a direct sales force of a certain size, an extensive training plan, an outbound telemarketing department for lead generation, a corporate pricing plan, and flexible payment terms for slow-paying (but sure-paying) corporate accounting departments. A goal to reach a specified penetration of the consumer market might specify a strategy to use retailers as commissioned dealers, an extensive cooperative advertising plan, an inbound telemarketing operation that can close as well as qualify, a tight credit and payment policy, and a special rate plan.

The corporate goals deal mostly with the parameters of profit and growth. The subordinate sales goals deal with revenue generation goals, but will have goals for costs also. The difference is that the corporate profit goals recognize corporate revenue and expense. Marketing goals encompass all corporate revenue (for the cellular operation) but only the marketing and sales portions of cost.

SAMPLE STATEMENTS

The Mission

To establish a cellular carrier operation that is grounded in the development of a service that derives its competitive advantage in superior equipment and design, superior customer service, and premium pricing to that segment of cellular customers who demand the highest service standards and perceive high, long-term value in cellular service. The company is thus focused on a long-term, major market presence using the highest quality equipment and personnel, while providing maximum financial return to the shareholders through customer loyalty.

Strategic Objectives

Maximize financial return as measured by cumulative gross operating income (i.e., net revenue less operating expenses) or Income Before Interest and Taxes. Cumulative measures in planning assure prioritization of long-term objectives and growth strategies over short-term opportunities.

The company will endeavor to concentrate on growth rather than cost savings as a source of long-term financial improvement.

The company will not forego investment and system growth in order to achieve short-term return on capital investment.

Strategies

1. Achieve long-term customer loyalty at premium prices.
2. Expand to additional markets through acquisition to increase growth.
3. Concentrate on a target market of highly mobile male business executives as outlined in the market forecast, whose needs and willingness to pay are consistent with the premium service offering, premium pricing, and quality customer service assumptions.
4. Establish internal sales channels for premium corporate customers, high-quality external agent channels for the general target market, and external alternate channels to test penetration of additional market segments quickly.

These are examples of such statements and are not intended to represent unilateral strategies that are superior to any others. The importance of the examples is to show how strategies fit with objectives and complement each other. At this level, the actual quantitative goals are neither specific nor time-dependent.

For the sake of brevity, the strategies and assumptions below deal mainly with pricing but are based on a complementary relationship with other marketing disciplines. Similar strategies and assumptions should also be developed for product, promotion, and distribution. As each set is developed, they should be compared for complementary relationships with the other disciplines and modified until they are cohesive, integrated, and consistent.

MARKETING OBJECTIVES AND PRICING

1. Achieve a 1.5% penetration of the market population over three years.
2. Achieve revenues over the same period equivalent to $80/month/customer or greater.
3. Maintain marketing expenses less than $400 per new customer.

Strategies to Achieve Marketing Objectives

1. Establish an image of quality differentiated from competing carriers through exceptional customer service, quality engineering, and premium equipment.
2. Do not use low price as a market entry vehicle; concentrate on superior value. Create a premium price strategy consistent with an image of premium quality.
3. Establish premium price levels with price structures that appeal to several main market segments, without appearing to "cream-skim" the market by pricing too high.

4. Establish retail channels that provide quality service delivery to the end user, and whose quality can be strictly controlled.

5. Establish alternative channels to penetrate price-sensitive and marginal market segments at lower marketing costs.

Business Assumptions for Pricing

1. The company will concentrate on cellular service and not attempt to make equipment and accessories a major business or profit center in competition with cellular service or the success of other channels.

2. The company will utilize a mix of sales channels to assure efficient delivery of service to all customers, to create opportunities for other local businesses in cellular in partnership with us, and to penetrate uncertain target market segments quickly through the following:

   - A direct sales force for major corporations who need long-term decision support and follow-on support.
   - Exclusive sales agents to provide high-quality sales and service, one-stop shopping, and professional, local customer service.
   - Resellers buying at wholesale prices to provide more rapid growth in net income, lower average marketing cost, broader market penetration, and broader pricing and service options than two carriers can provide.

   In order to best control the quality of retail delivery, the company will target its retail internal and agent channels to assume approximately 80% of the retail sales capacity of the carrier operation, assuming that resellers and independent dealers can deliver the other 20%.

3. Prices may vary from one market to another depending upon demand, need for service among target segment population, size of coverage area, and similar factors.

4. Bulk service at wholesale to resellers will always be lower than or equal to the lowest retail prices to the larg-

est end users, in order to provide a large opportunity for alternative retail providers while maintaining premium retail prices as a price "umbrella."

5. The pricing structure will be simple and targeted to attract the highest-potential market segments.

6. Retail service through internal and agent channels will be branded with the highly-visible corporate logo and premium branding. The wholesale service shall carry the name of the company, at the option of the reseller, but not its logo. Internal channels and agents will have exclusive use of the logo for advertising, branding of equipment, and service identification.

7. Pricing will be based on customer willingness to pay and perceived value, as constrained by financial return criteria, rather than target rate of return or cost-plus methodologies.

8. The carrier competitor cannot bear the financial cost to price more than 25% below us. Even if price differentials are this great, our superior system and service will be competitive to the target markets.

Specific Pricing Structure Assumptions

1. Regulators will permit premium, market-based pricing and high financial returns.

2. Pricing structures that include premium services in basic rates are most attractive to the target market (e.g., local toll included in airtime, detailed billing and vertical services included in access).

3. There are no known economies of scale or system equipment price reductions that can be used to assume wholesale and retail price reductions over time, even under competitive pressure.

4. Cellular service has a very low price elasticity for the target markets that have a high need. This means that there is no increase in demand when prices are lowered such that total revenue is increased. Price structures for target markets should always be priced as high as possible.

Lower prices, which attempt to attract new customer segments that are more price-sensitive than core target customer segments, should always be promotional rather than permanent, and not be crafted to attract existing customers from other plans.

Complementary Marketing Strategies and Assumptions

1. Strong promotion will provide adequate penetration of the target market without discounting price.
2. Discount pricing as a market entry/penetration vehicle is not appropriate, given other marketing strategies for the following reasons:

   - It does not fit with a premium service image.
   - The target market is more sensitive to quality than price.
   - Effective promotion and sales incentives should better achieve sales objectives. Low pricing will leave money on the table if these measures are effective.

These are examples of how higher-level objectives and strategies provide the basis for more detailed ones at the tactical level. While these have concentrated on pricing, they affect all aspects of the marketing mix and strategies for product, promotion, and distribution. We will discuss actual pricing in another chapter.

These particular objectives and strategies might actually be useful to some carriers but are merely illustrations of how such strategies and objectives must be explicit, consistent among elements of the marketing mix, and measurable. Without such explicit statements, managers often are left with broad, over-general strategies to achieve "quality." Without the guidance they need, they might contract sales channels without regard to service quality and market segment concentration or produce pricing plans that do not fit the overall strategy.

There is much information available today that might lead cellular players to entirely different marketing and pricing strategies: the emergence of the consumer market, limited capacity between the two carriers in some markets, and other influences. The key is to ensure that all objectives, and the strategies used to realize them, reflect the same information and assumptions, and that product, price, promotion, and distribution strategies all complement each other in the achievement of the objectives.

The next layer of such a strategic plan would include 1-year and 3-year goals for revenue, cumulative operating income, and the customer forecasts (usage assumptions, sales force sizing, price levels, etc.) that support them. Additional portions of the plan would outline assumptions, objectives, and strategies for each functional area (finance, billing, information systems, engineering, operations and maintenance, human resources, etc.). For a decentralized company, each region or market would adhere to the basic assumptions and strategies but include different objectives and forecasts (which quantitatively add up to yield the corporate projections), areas of concentration, programs, and tactics.

The 3-year strategic plan differs from the 1-year plan because it includes areas of development with a horizon greater than one year: the growth of a new sales channel, the cultivation of a new market segment, addition of new products, technology changeovers, and such. It has more uncertainty in its forecast and less detail in its determination of tactics and resources to accomplish its goals. However, it provides the opportunity to extrapolate beyond the limited resources and constraints of the marketing plan to take more advantage of unrealized growth potential.

THE ANNUAL MARKETING PLAN

The 1-year plan constitutes the immediate operational plan, forecast, and budget for Marketing as an integral part of the 1-

year operations plan for the entire business. It lays out specific programs for the coming year and must detail specific programs to achieve quantitative results.

Therefore, the marketing plan must outline the following, for example, and relate them directly to quantitative forecasts and costs: the total forecast of new customer gain, the number from each market segment, what penetration of each targeted market segment is required, the advertising programs used, the media schedule and the number of sales leads generated, the number of sales in each channel, the commissions paid to each channel for each kind of customer, sales per salesperson, and gain in channel size required to achieve the forecast.

Assuming that the carrier has already been in operation for a year or more, the basis of the marketing plan is the expected revenue from the existing customer base, recognizing "churn," or customer attrition (Figure 4–1). Growth in the customer base should recognize the replacement of lost customers in the current base and losses occurring during the year, in addition to the growth required to achieve a forecasted end-of-year increased base. While the gross sales levels required should recognize the need to replace churn, sales should not be held responsible for actual net gain (gross customer gain less churn), but for achieving their gross gain objectives. Churn is not usually controllable by sales channels; however, it can be manipulated. If commissions are paid for customers who last less than 90-180 days, the introduction of a minimum subscriber life greater than 90 days will substantially reduce short-term churn.

The total revenue from new customers must take into account the concentration of new customers in different segments from the base, weighted by the varying revenue per customer of each of these segments. The revenue from new customers must be distributed throughout the year in concert with the timing of actual customer gain, month by month.

The bulk of the forecast should be underwritten by committed quotas from internal and external channels. Detailed programs for immediate acquisition of new channels and chan-

Subscribers—Beginning	40,000
Net gain objective—20%	8,000
Subscribers—Ending	48,000
Churn @ 3.3%/month	19,008
Gross gain required	27,008
Average customer base	44,000
Revenue	
Startup fees @ $25.00	$675,200
Monthly access (44,000 × 12)	
5,000 @ $25.00	$1,500,000
18,000 @ $50.00	$10,800,000
21,000 @ $40.00	$10,080,000
Peak usage (44,000 × 12)	
5,000 @ 50 min. × $.80	$2,400,000
18,000 @ 80 min. × $.40	$8,448,000
21,000 @ 120 min. × $.36	$10,886,400
Off-peak usage @ 20% of peak	
5,000 @ 10 min. × $.24	$144,000
18,000 @ 16 min. × $.24	$1,013,760
21,000 @ 24 min. × $.192	$1,161,216
Estimated Revenue	$47,108,576
Less 20% of revenue at wholesale, discounted 20% access/startup, 15% usage	$1,643,809
Total Revenue	$45,464,767

Figure 4–1 *Excerpt from marketing plan forecast*

THE ANNUAL MARKETING PLAN

Sources of New Customers	
Agent A quota @ 12 salespeople—25 sales/mo	3,600
Agent B quota @ 20 salespeople—25 sales/mo	6,000
New Agent C quota @ 8 salespeople—25 sales/mo	2,400
Major Accts quota @ 6 salespeople—20 sales/mo	1,440
Retailer A quota	4,000
Retailer B quota	3,000
Dealers at 10% over previous year	2,368
Resellers at current activity level	4,200
Total Gross Customer Gain	27,008
Leads required	
50% of agent/direct quotas	13,440
Cost @ $18.00/lead	$241,920
Commissions: Agents @ $350	$4,200,000
Major Accts @ $150	$216,000
Dealers @ $200	$473,600
Residuals @ 0.5% of retail startup/access/usage	$1,884,343

Figure 4–2 *Excerpt from marketing plan*

nel members (new agents, sales force and quota increases, etc.) should be documented in the plan where gains are not assured through quotas and commitments.

The forecast for new customer gain (Figure 4–2) must be reconcilable to the known productivity of salespeople, locations, sales force size, and/or stores. For direct salespeople, the number of qualified leads required should be quantified, and the advertising, promotions, and telemarketing programs and staffing required to generate them should be recognized. It should be shown that the highest-cost channels are producing customers in the highest-usage, longest-lived customer segments.

The total cost of the marketing and sales efforts can be quantified from the documentation above and recent experience with these expenses.

If substantial revenues are generated via wholesale business through uncontracted resellers, such revenue is more difficult to commit and forecast, other than through trending existing business. The uncertainty of such revenue is a risk. This can be minimized by documenting the continued marketing and account support programs for resellers in support of this portion of the forecast.

Sales and marketing resources must be separated between retail and wholesale. Ideally, a separate *pro forma* income statement should be prepared for wholesale and retail operations, or for each individual channel. This should show the higher retail revenues obtained at higher sales costs than the wholesale channel. The return on sales for each portion of the sales channel mix will demonstrate the need to emphasize the most profitable ones (see Cellular Sales Productivity in a later chapter).

The development of a credible plan based on projected revenues and costs as they are derived from detailed plans will provide a road map for the year's marketing operations and will be invaluable in the planning process. The more detail that can be developed and the more the participants themselves are involved in the plan, the more valuable it will be to the entire team.

CHAPTER 5

◆ PRICING

Pricing is one of the basic four Ps of the marketing mix, yet it is given very little attention. In addition to its importance in defining margins, it is strategically important in what it communicates to the customer relative to the other elements of the marketing mix. Unlike some popular conceptions of consumer "rip-offs," price is not something involuntarily exacted from an intimidated buyer, but an amount freely paid by consumers when their perception of value and benefit gained is much higher than the price stated.

Pricing communicates value in its simplest sense when it is lower than that of a competitor for a product or service of comparable worth. Yet it also communicates value when a distinctive product is priced at a premium. A product designed, for example, for high quality and distinctive features should be associated with a premium price. Such a product should not use low pricing as a market-entry or market-share-gain strategy. The advertising and promotional strategy should do that job in such a case, not low price. The distribution strategy should emphasize personal service and knowledgeable salespeople. An alternative strategy, such as pricing below the competition, would have different complementary strategies in the other areas of marketing.

PRICING DEVELOPMENT

Pricing, as used here, flows directly from the customer expectations of the value placed on the service through experience and market research, as the primary determination of price levels. Actual pricing levels and structure must fit with the strategic and marketing plans, in addition to any customer expectations. "Willingness-To-Pay" (WTP) is the overall amount potential customers are willing to pay in relation to the value they place on a service, and is divided into structural elements and quantities purchased (access, usage, features, etc.) according to customer need and other inputs to the pricing process. It is assumed that all of the pricing flows from the strategic plan and marketing plan objectives and strategies. Once the appropriate market research and planning have been done, the basic pricing strategy and levels should almost fall into place by themselves, side-by-side with complementary product, promotion, and distribution strategies.

The *last* step in market-based pricing is to determine whether prices determined in such a manner fit within financial, billing, and regulatory constraints. It is unfortunate that good pricing strategies are tainted by having been subjected to such constraints before considering customer need. On more than one occasion systems people have said, "We can't bill that way." The mere expression of such a thought is abhorrent in a world supposedly guided by customer need, but more so because of the insolence to believe that the possible is circumscribed by the existing.

It would be naive to think that the constraints placed on pricing (and marketing in general) by financial, billing, technical, and other areas are not formidable. They may severely influence pricing in its final form, and thereby affect forecast and growth as well. The key point is that these constraints should not be prior restraints on the marketing process. Any substantial changes in pricing levels and structure derived from these constraints require a complete recycling through the strategic and marketing plans, to revise the strategic position of price among the other marketing elements as it is modified by constraints.

PRICING AND THE MARKETING MIX

Pricing should have a complementary relationship with other parts of the marketing mix, despite its having many strategic and tactical aspects of its own. Pricing also includes all of the terms and conditions of sale: not only how much the service costs, but the way it is paid, the way it is billed, and the terms of delinquency and collection.

In cellular, the customer's focus on the telephone instrument rather than the intangible subscriber service has had interesting pricing implications. Customers have noted the rapid decline in cellular telephone prices over just a few years. For customers who do not have a high business productivity need for the service, there is an expectation that the price of the service will come down, merely because the price of the CPE has declined.

Actually, it is the rapid decline of telephone prices which has allowed service prices to stay up. Customers easily overcome what once was the major price hurdle of the telephone, and can control service charges by limiting access. Service is an afterthought to the equipment decision, and customers are not as sensitive to service price as they are to getting a good buy on the equipment. These observations are changing as more price-sensitive business and consumer prospects enter the market. Newer customers are experiencing sticker shock when they see their first bill, and canceling service in many cases.

The major purpose of designing service pricing structures is to use the balance of access and usage to provide value to target customers. The consumer segment wants low fixed charges, while corporations want plans that provide discounts representing the value of the business they bring to the carrier. By combining low fixed charges with high peak usage rates for "economy" plans, and providing discounts to market segments associated with high usage and tied to customer loyalty, the price structure provides the flexibility different segments require without putting great pressure on overall pricing levels.

PRICING STRUCTURE AND LEVELS

The actual development of pricing structures and levels is a key strategic function, and must use the information developed in the strategic and marketing plans, combined with other elements required of a pricing plan by customers, such as the following:

- Consistent with customer needs/benefits
- Willingness-To-Pay (WTP) levels from market research
- Simplicity
- Certainty
- Quantity
- Competitive Comparison

Pricing must provide the service levels a customer needs to achieve value at prices that meet the constraints of a total amount per month the customer is willing to pay. Prime target market segments for cellular in the very largest markets might need to achieve, for example, about 200 minutes of usage per month, concentrated 80% during peak usage hours, at about

PRICING STRUCTURE AND LEVELS 53

A. Total minutes per month usage	200
B. % Peak usage	80
C. Access as a % of total bill	30
D. Customer willingness-to-pay	$120
E. Off-peak price as a % of peak	60
F. Access price	$36
G. Peak usage price	$0.46
H. Off-peak usage price	$0.27

Calculations:

$F = C \times D$

$G = (D - F) / \{(B + [1-B] \times E) \times A\}$

$H = E \times G$

Figure 5–1 *Cellular pricing model*

$120. Assuming that the user places a value on being reachable, and on being able to place calls from anywhere, the access portion can be valued at about 30% of the $120 WTP. Assuming off-peak prices are about 60% of peak, access and usage prices can be directly arrived at from this information, as shown in Figure 5-1.

Assuming that you have similar information for any market, the same calculations can be used to derive preliminary basic pricing. This plan appeals to the target market and must be tested against financial goals and forecasts to be certain it achieves all objectives. In addition to the tests listed above, these other elements of the pricing plan must be met: Is it simple? Can the customer estimate and calculate the total cost with reasonable certainty? Does it provide incentives for quantity purchase and usage? Does the plan invite favorable competitive comparison?

Multiple Pricing Plans

Additional pricing plans must also be created for other target markets. There are four or five pricing plans, each of which should appeal to general market target segments and should be named to convey to the prospective user that the plan is designed to be beneficial for his or her type of usage. The five general plans usually offered are variants of those listed in Figure 5–2.

The Basic Plan, already described, appeals to the key target market—executives of small companies and professionals. The Executive Plan packages some amount of usage (usually about 100 minutes) with access, giving a discount of approximately 7% on the packaged minutes. It is important to discount the packaged minutes rather than the excess usage, so the subscriber perceives a fixed discount, and large amounts of incremental usage over and above the packaged amount are not necessarily discounted (unless desired).

The Corporate Plan provides an incentive for 15-20 or more users to be listed on one bill to one address at a corporation, and usually includes a larger amount of usage packaged with access than the Executive Plan (perhaps 200 minutes). There is a larger discount than the Executive Plan for this plan, because it encourages multiple users and reduces administrative costs. It may have an associated minimum term of one year and also may include vertical services (e.g., custom calling features or voice messaging).

The Economy Plan is specifically designed for consumers, price-sensitive customers, and those who are not sure about their commitment to cellular. Rather than a simple discount, its balance of access price to usage price is altered to change the relationship of fixed (access) costs of service to variable (usage) costs for the customer. The Economy Plan charges an access price approximately one-half of the normal rate, and a peak usage price approximately double the normal rate. Thus, the user has the ability to keep charges low by using the service at a minimum and by calling off-peak as much as possible, if he/she has the flexibility to do so, as many consumers do. However, the Economy Plan is named specifically to attract price-sensitive customers. It doesn't *save* them money at all levels of usage, but it *allows* them to save money if they manage usage.

Comparative Plan Features

There are many variations on these plans, but they have several elements in common. In addition to targeting specific segments, some include discounts and free benefits which recognize the quantity of business that the customer contributes. Others provide these discounts and benefits in return for a minimum-length subscription commitment, usually one year.

The names used are not merely attempts to distinguish one from another. The names should be carefully chosen to suggest the type of user for whom it is best suited or suggest the benefits of the plan. This is important to help consumers come to a decision, to assure them they are on the plan that provides the most value, and to help position the service as the salesperson presents it to the sales prospect.

Each plan may have an alternative name and a slightly different structure, but there should be a distinctive plan incorporating these themes to appeal to major market segments. It is important not to have too many plans. They should have names that connote to the user the value they receive, or the positioning with a certain style or status (nobody wants to be known as a "heavy user"). It is natural and normal for some users to select a plan with minimum usage associated with it, even though they may not use the allotment every month.

Pricing Pitfalls

It is counter-productive to have too many plans, plans with confusing or non-meaningful names (e.g., Plan A, B, C, ... J), or to switch customers from one plan to another by calculating their bill according to each plan and switching them to the least

Basic Plan:	Access $36.00, Usage $0.46 peak/$0.27 off-peak
Executive Plan:	100 minutes included, 7% discount, 1-year contract, excess Usage at Basic rates
Corporate Plan:	200 minutes included, 10% discount, 1-year contract, excess Usage at Basic rates
Economy Plan:	Access $20.00, Usage $0.72 peak/$0.27 off-peak
Wholesale:	Basic Plan discounted 25% on Access, 15% on Usage (100 lines, 10,000 minutes/month minimum)

Figure 5-2 *Illustrative cellular rate plans*

expensive. Such strategies confuse the customer, complicate billing, leave money on the table, and most importantly forget the reason for multiple plans—to enable the customer to make a good choice easily.

It is sometimes tempting to lay out price structures that may attract customers by deceptive packaging. However, in any subscriber service, the customer will soon learn from the monthly billing if it is different from expectations or if the wrong usage plan has been selected. Any short-term revenue advantage of a deceptive plan will be far outweighed by long-term customer dissatisfaction. In addition, many marketers have learned that consumers are incredibly adept at seeing through pricing strategies and getting the best of the vendor.

Giving away voice messaging free with cellular service, for example, in the hopes of increasing airtime use through retrieval of messages using cellular service, will often result in customers using cellular voice messaging in lieu of voice messaging from the telephone company. They will use it to collect unanswered calls from all of their telephones, accessing it free from landline telephones to retrieve messages. This is far removed from the carrier's intentions. In many cases, however, some additional cellular use will result from the offer, and the customers' loyalty is increased because they feel they are craftily using the carrier's services to their own advantage.

Thus, it is important in pricing to understand that the cellular carrier should never expect customers to behave in the way that the carrier would desire regarding pricing decisions. On the contrary, they should always expect customers to behave in the way most self-serving to the customer and detrimental to carrier profitability.

Revenue Relationships

The plans should be designed rationally so that at each level of use, the appropriate plan is the best value for the customer (Figure 5–3). Below 60 minutes of use, the Economy Plan should be the best value; at 100 minutes, the Executive Plan; at 200 minutes, the Corporate Plan. The Basic Plan is priced to be a better value than the Economy Plan at any level of usage higher than about 60 minutes, but will always be higher than any plans that require minimum usage or a contract length. If plans are not designed carefully, users willing to pay high amounts for usage will migrate to other plans that are priced lower to attract different types of users.

Wholesale Pricing

Because some carriers do not see the value of resellers, they tend to structure wholesale prices (if they have them at all) as small discounts from each element of each retail pricing package. But the reseller plays an important role in adding diversity to the marketplace by providing its own set of retail prices. Wholesale prices should be structured after the Basic Plan only, but priced lower than any retail plan based on a much larger minimum quantity (e.g., 100 access lines) and lower sales and support costs. This gives the reseller the flexibility to lay out retail pricing structures different from the carrier. Wholesale pricing derived from direct discounts on each retail plan invite the reseller to use copycat plans and allow almost no room to do anything else. This limits the diversity of attractive features that different providers can use to attract different kinds of customers in the marketplace.

Also, it should be understood that some large companies may also qualify for such rates as end users if they are willing to accept the lower customer service and billing constraints of the plan. The wholesale plan requires certain restrictive features in order to allow the carrier to price it so low. The reseller (or large multi-user account) pays for access in advance and is billed for a minimum amount of usage in advance; receives only one bill, via tape; has a very short payment window; can have its customers assumed by the carrier for non-payment; has limited customer service, not accessible by end users; and may have other restrictions.

With these restrictions, the carriers's sales, float, billing, and customer service costs are kept very low. The wholesale business actually can be more profitable than the retail business, while affording a great opportunity for independent businesses in the community and diversity of service choices for the customer.

Crossover Point

Referring to Figure 5-3, the exact number of minutes at which two plans are equivalent, a "crossover point," can be calculated and adjusted based on minutes of use of the segments attracted to each plan. The average number of minutes per user attracted to a particular plan should lie midway between the crossover point with the plan for lower-usage and higher-usage customers.

More or different plans can be created. Marketing intelligence should be able to identify groups of potential customers with different needs who might be attracted to different levels or plans. In all cases, the plan intended for a target group should be the most economical at the average or modal level of usage for the group, and only at that level. The Basic Plan is undifferentiated, and charges under it will usually fall between

PRICING STRUCTURE AND LEVELS

	INCLUDED MINUTES	MONTHLY ACCESS	USAGE		******MINUTES OF USE******			
			Peak	Off-Peak	50	100	200	300
Basic	0	$36.00	$0.46	$0.27	$57.10	$78.20	$120.40	$162.60
Executive	100	76.26	0.46	0.27	76.26	76.26	118.46	160.66
Corporate	200	108.36	0.46	0.27	108.36	108.36	108.36	150.56
Economy	0	20.00	0.72	0.27	51.50	83.00	146.00	209.00
Wholesale	0	27.00	0.39	0.23	44.94	62.87	98.74	134.61

Figure 5–3 *Comparison of illustrative rate plans*

the Economy and the volume discount plans. The Wholesale Plan should be substantially lower than any retail plan at any usage level.

Pricing Plans and Sales Commisions

When we look at sales channels, we will investigate another important aspect of the rationalization of pricing plans across market segments. Carriers are discovering that as markets start to mature and attract marginal customers—those who use the system less because their need is less, like consumers—that sales commissions and incentives need to be based on the revenue and longevity potential of the customer. Since individual customers are difficult to classify, it is useful to allow customers to identify their own segment, and thus their usage potential and longevity, based on the rate plan they choose, and for the carrier or reseller to base sales incentives and commissions on the rate plan thus chosen.

This is enormously useful in order to match sales costs to revenues, to compensate salespeople and channels appropriately, and to ensure that salespeople are signing only customers who belong to the market segments to which they are directed to sell. However, it places an even bigger burden on the pricing plans to be well-structured, attractive to the right users, and positioned correctly relative to each other. If misused, and salespeople are, for example, overcompensated for attracting customers to the Executive Plan and undercompensated for the Consumer Plan, a lot of light users will find salespeople pressuring them to choose an inappropriate plan.

VERTICAL FEATURES AND OPTIONS

Vertical features (call forwarding, three-way calling, etc.) have very low penetration in cellular. Most calling is outgoing, so very few users have a need for these services. Salespeople view selling such services as a distraction from the sales presentation and a possible deal-killer if they stay to long after the close to bother with such details. Most are priced at about $1.50/month, but usually they are given away with premium pricing plans. The penetration of voice messaging averages 9-15% and can be sold at $5-10 per month, if incentives are provided, which also increases usage. Voice messaging is also packaged free with premium rate plans.

ROAMING

The use of cellular phones in cities other than the subscriber's home city has been an enormous revenue boon for cellular carriers, especially in larger markets and in those smaller markets immediately adjacent to large cities. Larger markets may realize 9-10% of their revenue through incoming roamers, most of which goes directly to the bottom line, unlike the high cost of obtaining new subscribers. Roamer revenue as a proportion of total revenue is growing rapidly with the high sales rate of portables, which enable business travelers to make cellular part of their air travel plans, and improvements in technology, ease-of-use, billing, and fraud control. Markets adjacent to large metropolitan areas may hardly bother to gain subscribers, but merely bill roamer usage of metropolitan users. Roaming has several important pricing considerations, even though fraud is the biggest issue concerning it.

Since cellular calling is mostly outgoing, many users can get along without being easily accessible in a foreign city, although technology is allowing incoming calls to roamers to be passed to them. Roaming is growing rapidly, and soon customers will routinely expect to be reachable in any city. Some crafty cellular customers, however, subscribe to cellular where monthly access is the lowest, giving them the right to use service in major cities, where access rates are higher. This, plus the premium that users place on the value of roaming, should put the roaming usage price at about double that of premium airtime at basic rates.

In addition, because of the cost of clearing roamer billing to the home carrier and the fact that many callers only place one or two calls per visit to the foreign city, many carriers rightly place a fixed daily access/registration charge on roaming. This only becomes a headache to users if they drive through several cities and make phone calls from each city in one day; however, it also encourages roamer fraud. With increasing automation of roamer billing and fraud control, such registration charges may go away eventually.

FINANCIAL CONSIDERATIONS

We have been considering the ways in which pricing must be structured in order to be consistent with market needs, attractive to particular kinds of customers, and complementary to the other elements of marketing. Our example of the pricing model uses particular prices that are only pertinent to a single, hypothetical cellular market, and should not be generalized to any or all real markets. Every market deserves its own pricing structure and levels. Pricing at the same level and structure in multiple markets not only implies ignorance of market needs and pricing strategy, but it leaves money on the table. It demonstrates that the carrier values ease of price setting over profit and growth.

Relationships of Pricing Elements

The major consideration of pricing for financial purposes is the price level of one minute of peak usage. The start-up fee occurs only once, and regardless of its level is not a major element of the revenue contribution of customers when considered over the life of their subscription. The monthly access fee contributes significantly to this figure and becomes more important as monthly usage per subscriber in marginal market segments declines. Off-peak pricing does not stimulate users to move from peak to off-peak usage, but it attracts non-business users and special segments to the service and implies the greater value of peak usage at higher prices. It provides lower revenue because of both its lower price level and the lower usage that occurs off-peak.

The peak usage minute price therefore determines, more than any other element of price, the total revenue and the marginal profit of the carrier. The marginal profit of a one cent increase in peak-usage price is almost equal to the entire large revenue gain achieved. The only negatives to raising the peak usage price are the loss of price-sensitive customers and its competitive and regulatory implications. A simple financial model will illustrate the sensitivity of revenue and profit to peak usage price over other pricing elements.

Price Elasticity

Within this context, it is surprising how little empirical experimentation has been done to raise prices in the cellular environment. Most efforts have been aimed at lowering access and usage prices to stimulate customer growth and to provide discounts as an incentive to loyalty. The core target market for cellular was predicted through market research, and has demonstrated, to have little sensitivity to price. This is confirmed by the lack of usage and subscriber growth through price reduction. The general statement is that cellular has low price elasticity. Reduction in price will not increase total revenue through increased customers and usage. As a corollary, *the top price that core users will pay for peak usage without a reduction in total revenue has never been tested.*

There have been some examples in which service price reductions have stimulated growth. Such growth, however, is only within price-sensitive market segments like consumers. A general price decrease gives away money to existing heavy users and stimulates growth in the lowest-usage, most price-sensitive marginal segments, at a huge cost in overall revenue. Therefore, price reductions must be carefully targeted to affect only new users in segments that are price-sensitive.

Price Level Strategies

The price level of peak usage has changed little since the introduction of cellular. With some variation, the top 20 markets are priced at $0.35 to $0.45 per minute, the next 20 at $0.30 to $0.40 per minute, and others at $0.20 to $0.30 per minute. While the range of these variations alone should point to the ability to raise prices a few cents in many markets, the concentration on subscriber growth has minimized attention to this opportunity. Indeed, once two competitive carriers have settled into a market, each is rightly fearful of the potential loss of customers and reduction in relative subscriber growth rate from raising prices relative to their competitor even one cent.

Certainly one symptom that a price increase is in order is customer growth so healthy that the capacity of the system is strained, and service availability and system performance measures indicate usage congestion. Continued rapid growth during a period of congestion means that the service is underpriced for the value received or that sales incentives are too high.

Pricing action is an effective way to increase returns and prevent further service deterioration to current customers while system capacity is being addressed. Aggressive growth during such a period is dangerous to long-term customer satisfaction and retention.

While carriers should look at the possibility of the financial benefits of a price increase versus stimulating customer growth, their hesitancy to do so should also point to a need to carefully consider discounts and price reductions as a strategy to increase growth rates, customer loyalty, and reduce churn. A few generalizations might be helpful.

1. Price reductions in usage or access have never been successful in increasing total revenue. With the emergence of price-sensitive segments, price reductions should affect only those potential customers.

2. The Basic Plan serves as a "price umbrella," meaning that it is a benchmark for the levels of retail and wholesale pricing upon which all other plans in the market are based. It can be raised, but it must never be lowered, or it will trigger proportionate reductions in all plans in the market offered by all competitors.

3. Any price reduction deemed to be required to attract a new market segment should be carefully limited to a price plan that affects *only* new users and *only* the segment to be targeted. General price reductions merely give away money to new and existing users indiscriminately.

4. Any price change should be tested on a promotional basis in a single market before being permanently implemented in multiple markets.

The listing of these strategies should not imply that any price reduction is ever recommended. The continued focus of new customers on telephones rather than cellular service, combined with the monthly introduction of new cellular telephones with improved capability at reduced prices, favor promotional and channel strategies completely divorced from service pricing strategies for attracting new customers. For existing customers, evidence of price elasticity in particular segments is the prime factor driving price reduction, even in the face of competitive price reductions. So called "free minutes" promotions are usually only helpful in trading existing customers temporarily between providers, at high advertising and commission costs.

There is no substitute for the continuous evaluation of various scenarios of system growth, market growth, subscriber growth, price changes, and cost changes, through a financial model, as part of the ongoing overall strategic planning process. Coupled with a good reflection of historical performance of marketing strategies and other operating variables and programs, the process should demonstrate how price changes differ from market growth and cost reduction in improving financial performance. It will help providers discover the kind of stimulation in customer growth required to justify a price reduction, especially when the other marketing costs of attracting marginal customers are considered.

Long-Term Outlook for Pricing

Except for short-term adjustments to obtain the full potential of peak usage pricing, there is probably no opportunity in the long-term future to raise cellular prices further for voice applications. Revenue increases will come from additional services and more efficient use of spectrum. Nominal prices for basic voice communication will be pressured downward by the need to fill new capacity and by competition from future wireless services.

From a financial standpoint, payback periods and cumulative discounted cash flows are severely penalized if system capacity anticipates, and capital investment in system expansion substantially precedes, new customer growth. Still, competition and customer satisfaction require carriers to avoid system congestion at all costs.

Finally, financial modeling of pricing should assume that any generation of system technology will be replaced every five years. Five years ago the assumption could have been 7-10 years or more.

CHAPTER 6

SALES CHANNELS

Distribution is one of the four basic elements of the marketing mix. Delivery, or physical distribution, of an intangible subscriber service like cellular is really not applicable, so we will concentrate on sales channels as the more important aspect of distribution for cellular service.

AGENTS

As we mentioned in the introduction to the book, the authorized sales agent first emerged as the primary sales channel for cellular because of the need for technical radio and installation expertise from the existing radio industry, plus a need for "one-stop shopping" for cellular telephones, installation, and service subscription. In order to begin selling service, the use of existing sales channels for mobile telecommunications services enabled faster ramp-up of the sales capability as commercial service began. In addition, the importance of using local talent to participate in the market, and the flexibility of paying only for sales performance as a substitute for the bureaucracy and high cost of internal channels, cannot be overstated.

The definition of a cellular agent, as used here, is an independent business under exclusive contract to a cellular service provider (reseller or carrier) to sell cellular service at retail to end users in the name of the provider, in return for a sales commission. The term exclusive refers to the contractual obligation of the agent to sell only for the provider and no other in any one market. An agent is selected by the provider and is under firm contractual control of the provider with regard to location, services available, sales methods, quota, and quality standards. It usually is a business dedicated to cellular sales as a contractual obligation.

DEALERS

While a cellular dealer may be synonymous with an agent in some markets, a dealer, as used here, is an independent business that sells cellular service at retail in return for a commission. To distinguish dealers from agents, dealers have few contractual obligations to the provider, are not necessarily exclusive, and may change provider loyalties frequently. They may be contracted to an agent (see Broker-Agents, following) or reseller. Dealers may also have a contract with the carrier, which usually involves less control than the agent contract, no use of the carrier logo, and lower commissions. Such dealers are called "direct dealers", as distinguished from dealers who are indirectly connected to the carrier through intermediaries.

Dealers are useful also where the method of doing business may not match the carrier's or reseller's mainstream sales methods and agent contract requirements. Automobile dealers, for example, cannot meet the strictly controlled terms of an agent contract that may require a direct sales force or a technical support capability. While the carrier and automobile dealer may earnestly want to cooperate to provide cellular service, telephones, and installation, the auto dealer cannot be directed on how to sell or where to locate.

Dealer status allows existing companies to participate in cellular whose primary method of doing business is not congruent with cellular agency. Another example would be a retailer, such as a department store or consumer electronics store, which cannot change its method of operation to meet the requirements of an agent, yet can target the consumer segment much better than the cellular provider's mainstream sales effort. Such channels may have exclusive contracts with providers (direct dealers), but cannot operate under the same terms as an agent. Others may be one-person cellular sales entities with no contractual obligations.

BROKER-AGENTS

Broker-agents, sometimes called master agents or aggregators, have evolved as an additional step in the distribution chain. In addition to their contractual obligations as agents, broker-agents may serve as a clearing house for independent dealers. The broker-agent receives cellular subscriber agreements from independent dealers and its own contracted dealers, and supplies them to the provider along with sales resulting from its own efforts. The broker-agent services dealers with cellular telephones, installation facilities, and customer service support and pays them a portion of its commissions from the provider.

This relationship provides the independent dealer with faster, local service in large markets, more rapid payment of commissions, and other support that the provider may be slow or hesitant to provide to smaller dealers. Since commissions are paid to agents by the carrier as a combination of sales commissions plus residuals on end-user revenue, the broker-agent may pay the dealer most or all of the sales commission (or even a premium over it) in order to retain the residual commission for a more rapidly growing customer count.

The provider gains customers faster in this relationship through an arms-length arrangement with dealers who are too small to deal with directly or may not have the quality standards or desire for allegiance of the contracted agent. However, the provider in this arrangement has less control over the quality of the customer, the professionalism of the sales presentation, or the customer service provided to the end user by the dealer.

RESELLERS

Resellers emerged rather accidentally through the requirement of the FCC for resale. Various cellular carriers have reluctantly allowed the participation of resellers, because of their desire to control end users at retail prices. However, in many markets, resellers are seen as an important part of the distribution channel mix. They may attract certain market segments because of their customer base in other products; they may have a nationally recognized name and an existing customer base in other lines of business; and they may add diversity in sales approach, pricing, or other elements of the marketing mix that the carrier or other providers do not. Most importantly, they may provide greater profit to the carrier where the wholesale discount on the carrier's retail prices is less than the carrier's cost to sell at retail.

Many carriers want all customers at retail. Carriers can be uncomfortable that some resellers are carrier competitors or are selling for both carriers at the same time. The carrier may be concerned that the reseller will discount the retail price and erode the overall retail price level, or "price umbrella," for the entire marketplace. Carriers apply pressure to resellers by refusing to supply any marketing support and reducing margins between retail and wholesale. They can tighten credit terms and force resellers who don't pay on time out of business.

The arguments for using resellers still apply. Wholesale is often more profitable than retail, because sales cost savings can far exceed wholesale price discounts, and the carrier can set them based on sales cost savings. There are many more strategic and less direct reasons for encouraging resale. Introducing new channels of whatever type (profitably) is important, for two reasons. First, new segments can be approached using methods that existing channels don't (e.g., by offering alternative promotions, price plans, or levels of customer service). Second, channels must continually expand or the carrier will stop growing due to churn.

RETAILERS

Retail stores may participate in cellular as dealers, as described above, or as a special type of agent or reseller. Because of their increasing role in cellular as the consumer market grows, they should be treated as a separate category of sales channels.

Retailers become an important factor in a particular market as a sales channel because of three factors.

1. The rapidly declining effective price of cellular phones, which is the primary barrier to consumer adoption of cellular.
2. The increasing consumer knowledge of the benefits and use of cellular through word-of-mouth, without the aid of a direct sales force or heavy advertising effort.
3. The need to match the lower average monthly cellular revenue and shorter customer life of the consumer market segment with a lower-cost marketing and sales operation.

While it is expected that retailers will begin to play a large role in cellular distribution, experience has not caught up to forecasts. The emerging consumer market expects to find cellular phones in consumer electronics stores, yet many consumers buy from traditional agents. Retail consumer channels account for only about 5-10% of sales, although this number should grow rapidly. Based on the lower requirements for service and sales effort, it would be logical to assume that commissions for retailers would be lower than for traditional agents.

But while carriers are paying traditional agents lower commission rates for customers who choose consumer-targeted rate plans (see Pricing), they are paying comparable rates to retailers as agents and dealers. If the retailer truly is bringing in a lower-quality customer in the consumer, commissions should match the lower revenue potential. However, competition among carriers for the two or three largest consumer electronics retailers in a metropolitan area shoots commission rates up.

The payment of standard commissions to attract a less desirable segment may be a necessary short-term strategy to capture these agents and dealers channels for the long run. However, no individual agents or dealers should be paid rates that deviate higher from established rates for levels of performance. This would result in the requirement for all members of the channel to be paid the higher rate, without fail.

Retailers have the highest churn of any channel. They have the most consumers—the lowest-need, shortest-lived, lowest-usage segment in cellular. In addition, they usually do not want to have residuals in their contract, so they have no incentive to keep customers.

National retailers like Sears, Circuit City, and Radio Shack will insist on working directly with the carrier merely to assert their importance, even if other channels offer good terms. National retailers have the clout to get even better prices than the carrier on phones. Local and regional retailers, even if large, will use whatever provider pays the highest commission (usually without residual) and provides low-priced phones, demonstration lines with free usage in stores, and advertising co-op money. In either case, standard agent contracts do not work.

The carrier will normally provide full training to retailers. The problem is that the retailers don't want it. They like to obtain the information from carriers at national headquarters and distribute a memo to all their stores. If the carrier offers local training, either no one will attend or the ones trained will soon be moved to another department or store.

INTERNAL DIRECT SALES FORCE/MAJOR ACCOUNTS

The carrier organizes its own general direct sales force separately from its major accounts direct sales force. The direct sales force normally does not specialize based on industry, but may form several geographic branch offices with installation facilities or use agents or outside contractors to do installations. This sales force is formed to compete in the same market segments as agents and resellers—small business and general business—with a concentration in professional services, construction, real estate, and financial services. They are created to take business away from resellers and agents at a lower cost or higher revenue, so they create channel conflict rather than resolve it.

In addition to the cost issue, the carrier wants to control accounts directly if it suspects that agents and dealers are churning customers among carriers and resellers. Some carriers say they have shown that when a direct sales force controls customers from the same segment, they stay longer and have higher usage than agent-generated customers.

A key account for a carrier is a large corporation or "major account." These usually include the local offices of the Fortune 500, local firms of large size (over 1,000 employees or $100M in annual revenue), and government accounts. A "key account" for an agent, on the other hand, is any account which has 10 or more users. They are an important source of referrals and additional business. Easier administrative procedures such as special billing and usage reports are important things to offer major customers, but special pricing plans are the most important.

The major accounts group is usually broken down by industry within the companies that fit the definition given above. Each representative specializes in an industry or a few large companies and does multiple presentations to demonstrate applications pertinent to the company's industry that increase productivity through cellular. Large companies often require three to five presentations before they become an account. However, once the accounts are established, the repeat business is sometimes so good that carriers have to take the accounts away from the salespeople because they make too much money with little sales effort, ignoring the other accounts in their industry specialty.

Large companies were slow to implement cellular on a company-wide basis, but by 1987 they began to adopt cellular in a big way. While the carriers offered easy administrative procedures for subscription and installation, they also offered phones at discount prices, but above wholesale cost. Because of the rapid decline in phone prices, after awhile this was no longer an important advantage.

Today, price discounts on service are important. In California, however, special pricing different from published tariffs is permitted only for government agencies, where the price is sometimes discounted 50% from retail.

Because commissions to agents are an explicit sales cost, when commissions climb, the carrier tends to position its direct sales force in competition with its agents. Considering the carrier's overhead, it is almost always better to use external channels instead of an internal sales force. The exceptions are major accounts and direct marketing to consumers.

DIRECT MARKETING

Direct marketing is the ability to generate a sale directly as a result of a telephone call or mail solicitation, without a sales call or store visit. As awareness of cellular in the mass market increases, this channel becomes more and more viable. Because it requires a very carefully constructed promotion and a direct mail/telemarketing team, it is usually only within the power of the carrier to develop a viable program. The channel has great benefits for the carrier in reduced sales costs and commissions, direct control of the sales process and the customer, and increased satisfaction of the customer through rapid fulfillment of the sales transaction.

The carrier may merely offer a low-priced portable telephone or a free premium with the telephone and service subscription, through newspaper or radio advertising. Response is almost entirely by telephone, although printed advertisements usually offer a reply card also. The carrier arranges to deliver a programmed portable telephone by overnight express, with initial or permanent billing through a validated credit card transaction.

Direct marketing includes telemarketing and direct mail, where either results in the close of a sale. A distinct set of telemarketing talents are required in this environment. In addition to being a special case of these channels, direct marketing is also distinguished by its reliance on a feedback mechanism of productivity measurements and methodical working of a computer database of prospects. For this reason it is often called data-based marketing.

This particular channel will have increased importance, because portable sales will soon exceed 80% of new cellular telephone sales, and general consumer awareness of cellular is rapidly increasing. Like retail, however, this will not be a major channel for some time.

Telemarketing is an important part of direct marketing, and is further broken down into inbound and outbound. Each of these requires different skills and management. Outbound telemarketing is used as a direct solicitation of target markets, as well as for lead generation. Inbound telemarketing is a part of "pull" marketing, where advertising directs buyers to respond to the advertising by phone for fulfillment of the offer.

COMPANY STORES

Cellular carriers in the past have created their own full-service sales and installation centers with limited success. Such centers have difficulty in controlling costs to be competitive with independent dealers and retail agents. They also cause friction in the retail channels because the carrier is competing for the same customer as the agent. Since the carrier usually has only a few such centers, they do not constitute the bulk of retail distribution and do not become the leading channel in a market.

Unless there is a severe lack of potential agents and dealers in a market, company sales centers usually do more damage than good. They do not capture significant market share, they compete directly with other channels, and they cost more than the independent channels whose costs they are trying to cut.

The most recent efforts of carriers to directly control channels to the end users is the emergence of carrier-owned retail stores. Such retail locations have the same problems as their full-service center predecessors. The company store concept also entails the problem of establishing a new channel for a single retail product directed at consumers. Specialty stores are usually geared to customer needs and deal in an entire line of products for a specific purpose: pets, auto parts, furniture. Consumers tend to lump together consumer electronics and telecommunications, and expect to find these items at full-line electronics stores, not boutiques.

Thus the concept of a retail store for cellular only does not fit with the expectations of the consumer market. While it is important to have a storefront operation at agent and dealer locations to accommodate the consumer market, carrier-owned retail cellular specialty stores for consumers are not yet part of the success plan.

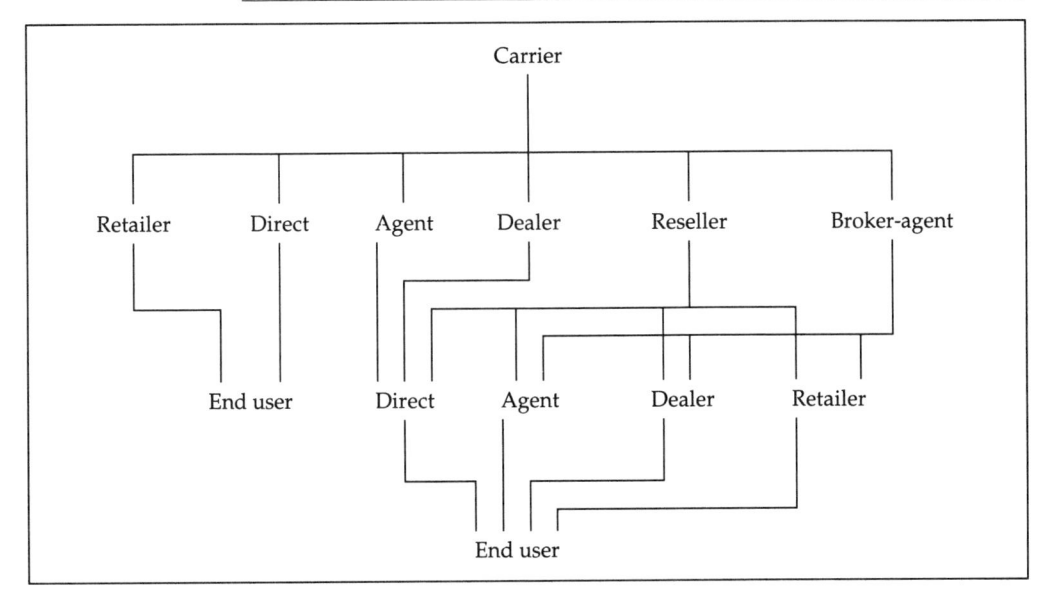

Figure 6-1 *Cellular sales channel relationships*

CHANNEL MIX

Figure 6-1 shows the various possible relationships of cellular distribution channels that are possible. Cellular can support a two- or three-tier sales channel arrangement because of the high number of small dealers and other small channels requiring local support. Also, the high commissions paid as a combination of up-front plus residuals provide enough incentives to share commissions through the channel. Where the relationship of wholesale to retail prices is adequate, usually in only a few of the largest markets, resellers can support their own independent sales channels below the carrier level.

The Hybrid Model

While some channel strategies may recommend themselves over others, the only true formula for success is to use a mix of channels, each of which concentrates on a different kind of customer. While there will be a broad overlap of segments that do business with each channel, the idea is to have channels that appeal to customers in different ways.

The carrier's direct sales force provides the kind of commitment and support that large corporations require. The traditional agents and dealers will find customers in the small business and consumer segments with the full complement of installation and technical help. Retailers target the mass consumer market. Resellers individually attract diverse markets such as ethnic groups, associations, factory-installed cellular units in new cars, or price-sensitive subgroups in many segments with price structures that differentiate or discount from the carrier.

MARKET SHARE BY CHANNEL

It is dangerous to attempt to portray a "typical" share of market by channels, but it is important for a carrier to understand the shares locally for both carriers. The figures below assume a market with reseller activity and an internal direct sales channel which handles only major accounts:

Agents and Dealers	60%
Resellers	20%
Direct sales	15%
Retailers	5%

SALES CHANNEL CHARACTERISTICS

Sales channels can be classified in order to distinguish their roles in the cellular marketplace (Figure 6–2). Because cellular is lucky enough to attract the interest of diverse interests as potential channels, it should cultivate all of them for current and

CATEGORY	INTERNAL/ DIRECT	AGENT	RESELLER	RETAILER
Target market	Large accounts/ government	Sm. and med. businesses	Sm. and med. businesses	Consumer
Compensation	Salary + low sales commission	High sales commission + residual; profit on phone sales	Margin; profit on phone sales	Sales commission; profit on phone sales
Branding	Sells in name of carrier	Sells in name of carrier	Sells in own name	Sells in name of carrier/self
Risk level	Low risk	Medium risk	High risk	Low risk
Return	Low return	Medium return	High return	Medium return
Investment	No investment	Medium investment	High investment	No investment

Figure 6–2   *Classifications of sales channels*

future use. Many cellular carriers have been less successful trying to control distribution through direct channels. They have failed in effectiveness because they couldn't cover the market, and failed in efficiency because their costs were too high. Rather than assume that they need to control customers at retail, carriers need to examine the tradeoffs of control with profit, growth, and other considerations to objectively determine the most appropriate mix of internal and independent channels.

By knowing the channel preferences of each target market segment, the cost per sale of each channel, and the customer volume each channel can deliver, the carrier can devise a hybrid channel strategy that most effectively and efficiently delivers the market. There is no single channel that can do both by itself.

There are other benefits to multiple sales channels. These include better coverage of market segments not attractive to mainstream agents and direct sales forces, multiple daily advertising presence in the media, and additional price competition on equipment to lower the "hurdle" price of cellular telephones as an entry barrier to consumers. One additional benefit is the ability to have enormous resources devoted to beating the bushes for cellular customers, with the external channels paid only for actual sales performance in new customers.

The agent, or exclusive dealer, has emerged as the predominant sales channel for cellular. The ability to sell a small business executive on a single direct sales call and to promise rapid activation, a quality mobile installation, and local, personal service on a flexible basis have all worked in favor of this channel. The carrier's own channel usually works quite well for major corporate accounts but is inflexible in the pricing, terms, sales approach, and rapid competitive response required to work most local markets effectively by itself. Agents and dealers make daily adjustments in phone prices, commissions, and incentives to make the most of any deal under fast-changing local conditions—a feat impossible to manage in a larger, market-wide organization. The position of the agent has been eroding, however, with the penetration of additional segments like large corporate accounts and consumer markets. Corporations require patience, staff work, repeat visits, professional presentations, volume discounts, and account servicing, which only the carrier can support through its major accounts sales force.

The consumer market, which expects to travel to a nationally known retailer to purchase a portable phone requiring no technical assistance or installation at a discount price, cannot support, and does not require, the kind of direct sales effort and services offered by the agent. If carriers and resellers are willing to pay the same commissions to a retailer that they are prepared to pay an agent, the retailer can discount the telephone at or below cost, subsidized by the commission (where permitted), because its sales costs are so low and spread over so many other products.

CHANNEL CONFLICT

Because of relatively high commissions paid for service activation by carriers and retailers to agents and dealers in lieu of non-existent profits on telephone sales, carriers and resellers may direct their sales forces to compete directly with their agents and dealers. Such practices do not recognize the implicit cost of internal direct sales forces (benefits, salaries, supervision, etc.), which, when added to lower explicit sales commissions paid, make the total cost of internal sales forces higher than external channels. The internal sales force should be directed at specific market segments that do not interfere with the efforts of agents and dealers directed at the general marketplace and core market segments.

Carriers may also reserve leads generated by promotions or telemarketing for their own internal salespeople. This forces internal channels to be in direct competition with their agents, and reduces the effectiveness of agents. If unchecked, agent salespeople will eventually be directed mainly at difficult prospects and cold-calling, while more expensive and highly trained internal sales professionals will be directed at easier sales at much higher costs. Each channel needs to be supported toward success, and paid according to performance as directed toward specific market segments, or it will wither and die. Sales management should treat both internal and external sales channels alike, as partners in success.

SALES COMMISSIONS

Sales commissions for cellular subscription are higher than they might be if there were no competition and better profits on cellular telephones, which were intended to be a large profit center for agents and dealers. Competing dealers may help drive commissions downward, but competition among carriers and resellers for the best exclusive dealers drive them up more than other factors.

Sales commissions to agents are structured based on a cash sales commission at the time a subscriber is placed on the service, with an optional residual commission calculated as a percentage of access and usage revenue (excluding toll, which usually has no markup). Typical up-front commissions in the top 30 markets may range from a low of about $200 to a high of $600. The average is around $300-350.

Residuals

Residuals may range from 2-7% of revenue, depending on the market and the size of the agent's base of customers, and are paid for 3-5 years or the life of the customer.

Many carriers and resellers have resisted paying residuals even to their best agents, because they have difficulty in relating the payments rendered to actual services provided or results in longer retention of customers through such incentives. However, they have more recently recognized some of the following important benefits of residual commissions:

1. The residual becomes a major component of the agent's revenue stream after the agent has sold the first few thousand customers. It becomes the carrier's "insurance" that the agent will never leave a carrier for a competitor and lose future residuals on existing customers.
2. The residual supports the overhead costs of the agent through slow sales periods. As a one-product direct sales operation, agents are extremely vulnerable financially to changes in sales levels. This effect protects the carrier by ensuring a quality agent's ability to provide continuous quality service.
3. The residual provides an incentive to the agent to prevent customer loss and to secure high-usage, long-life quality subscribers.

The residual structure has caused the development of the broker-agent in large markets. As defined above, the broker-agent passes the sales commission from the carrier through to independent dealers or uses it to provide cellular telephones to small dealers or one-person sales shops at subsidized prices. The broker-agent makes money on the climbing residuals from a rapidly expanding customer base achieved through multiple dealers. This operation becomes a satellite support center for dealers, with much more flexible and local services than the carrier may provide—and faster payment of commissions in cash.

The combination of a commission structure that supports several tiers of distribution and subsidization of the price of the phone, combined with a focus of the customer on the telephone decision rather than the carrier decision, results in an environment in which the carrier alignment of the channel which sells the phone determines to which carrier the customer subscribes. Thus, the carrier needs to use residuals as a tool to gain and keep the loyalty of the agent, retailer, and broker-agent.

RESELLERS

As mentioned, the role of resellers arose with the requirement of wireline carriers to allow non-wireline carriers to sell wireline service until the second non-wireline system was in operation. Thus, many of the wireline carriers have viewed resale as a burden that furthers the interests of their future competitor. They have tried to contain resale margins in order to minimize the penetration of their competitor, and have provided only minimal support of resellers.

Many carriers have not recognized the benefits of selling at wholesale. The following benefits are in addition to its relieving the carrier of the sales costs of an internal sales force and commissions to dealers and agents, which alone may be higher than the margin they give to resellers:

1. Resellers pay in advance for access and usage, and may relinquish their entire customer base to the carrier if they default on payments.

2. The cost of float, collections, and bad debt are all virtually eliminated for the carrier.

3. Resellers and their sales channels provide all customer service to their end users, the carrier needs to provide customer service to only one contact at the reseller for system activation/deactivation of subscribers.

4. Resellers provide all retail billing. The carrier needs only to supply the reseller a tape of billing transactions on a timely basis to one address.

5. Resellers provide a different set of offerings, prices, promotions, and terms to customers than either carrier, and, as an alternative provider in their own name, make the local provision of cellular service appear more like an industry than a near-monopoly.

6. Resellers like GTE and Motorola bring a national reputation to a local market.

7. Resellers access different market segments in different ways from the single retail approach of the carrier. They may also have a dedicated customer base from an unrelated product line as potential cellular customers, which the carriers cannot target directly.

8. Resellers control customer acquisition, billing, collections, and bad debt carefully, and thus provide a better-than-average quality of customer because their business is entirely based on financial efficiency.

Because of their desire for retail control of their customers, and, for wireline carriers, the heritage of having their competing carrier as their largest wholesale customer, most carriers have discounted the value of resellers without examining the profit potential of encouraging and supporting them. It has been shown through analysis of documents filed with regulators that, for carriers in California, retail operations may be losing money or are barely profitable, while wholesale operations are the main source of profit for the carrier.[1]

---

[1] Car-Phone Rates Run Into Static. *San Francisco Chronicle,* May 8, 1989, p. C1.

AGENT CANDIDATES	RESELLER CANDIDATES
Wish to work under the carrier's name	Wish to work under their own name
Willing to work under carrier/reseller's rules	Want to develop their own rates and business terms
Have low risk tolerance	Have high risk tolerance
Work for commissions	Work for margin/float
Require $0.5M finances	Require $1.5M finances

Figure 6–3 *Characteristics of agent versus reseller candidates*

AGENTS VERSUS RESELLERS

Agents are primarily sales organizations that are interested in using the carrier's name, logo, and support to promote sales. They make money on retail markup of the cellular telephone and installation, plus commissions from the carrier for new subscribers. Resellers are organizations that want to sell in their own name. They require an efficient billing and collections process and more capital and financial wherewithal to get started (see Figure 6–3). These characteristics help distinguish independent companies as candidates for different kinds of channels.

NATIONAL CELLULAR PROGRAMS

Motorola and GTE are major companies that are national resellers. Other, smaller companies also are national resellers in major markets, and make it easy for customers to obtain service for any market from one provider.

McCaw is the only company with a retail national accounts program for major corporations. It is respected in the industry and offers low-priced telephones and custom billing. Rates are determined by local market discounts.

AGENTS: THE PRIMARY CHANNEL

Agent Organization

Agents are typically organized as shown in Figure 6–4. All salespeople are generalists, and none are confined to certain industries or account sizes. The entire business is devoted to cellular only. There are one or perhaps two office locations within a market, and agents usually operate in only one market.

Agent Sales and Marketing Activities

Qualified sales leads are generated 40% by carrier advertising and telemarketing, 20% by internal agent advertising and/or telemarketing, and 40% through referrals of existing customers and the salesperson's own networking. Referrals are most effective, with carrier advertising leads second in importance. The internal outbound telemarketing group uses lists from listing companies, which are coordinated with the carrier to avoid duplication.

The carrier is the primary advertiser for cellular in the marketplace and sometimes supplies sales leads to its agents. It reserves leads for major accounts, government, education,

	PERSONNEL	PERCENT
Sales	20	56
Telemarketing	5	14
Installation	3	8
Customer service	3	8
Management	3	8
Clerical	2	6
	36	100

Figure 6–4   *Typical organization of agents*

institutional, and associations for its own internal Major Accounts group. General leads are shared with agents and the internal sales force (if there is an internal sales force for other than major accounts) on a pro rata basis according to the size of the sales force or sales quota of each group. The carrier designs promotions in cooperation with its agents based on featured telephones, assuring that the manufacturer can supply them to agents at an agreed on price. The carrier supplies sales training, which the agent supplements. Most marketing brochures, forms, and leave-withs are designed by the carrier and supplied to the agent.

Agent advertising is usually confined to small newspaper ads in the business section promoting featured cellular phones. The carrier uses newspaper and radio, and TV in smaller markets where it is less expensive. The carrier provides co-op of media advertising based on recent sales levels of the agent, but must approve the advertising layout and copy.

A direct salesperson used to average three sales per week, or 12-15 per month. With greater customer awareness provided as markets reach maturity, and sharply lower telephone prices, a salesperson in a major market now can be asked to average 25 sales per month at $100 commission per sale. Their only other benefit is cellular telephones and service. The commission may not be constant, but may include the activation commission plus some portion of the markup obtained in the sale of the equipment. The sales cost per activation begins at about $250, and fully loaded runs about $400. Salespeople receive no salary but sometimes draw an advance on commissions. At this sales rate, agent direct salespeople earn about $2,400-3,000 per month.

Value-added services are sold by direct salespeople if at all, only at the time of sale of the telephone, because it is too expensive otherwise. Such services are not even mentioned unless a need is uncovered during the sales presentation. Salespeople like to keep the presentation simple and leave as soon as possible after a successful close. Paging is generally not sold by agents. Although there is an increasing strategic relationship among the services of wireless voice, paging, and voice messaging, paging has not penetrated cellular users more than about 20%. The primary markets of executives, managers, and consumers are not associated with paging, while other segments usually already have it when they move up to cellular.

CUSTOMERS

As mentioned, agents have a sales rate of 25 per salesperson per month, or about 500 for a sales force of 20. In a mature large market, the typical agent's customer base will level off at about 15,000. At this level, with a churn rate as low as 3% per month, the company will lose as many customers as it gains if it does not expand its sales force, increase its productivity, move into storefront sales to consumers, or act as a broker-agent.

The agent targets the small business category for 50% of sales. These customers are most likely to buy on a single sales call from a direct salesperson. Another 30% come from mid-size businesses and business users from large companies who do not use the corporate account. Consumers constitute most of the remaining 20%. Consumers do not prefer a direct sales channel, are hesitant to buy on a single sales call, are more price sensitive than business customers, and usually (80% of the time) purchase a portable, which does not provide installation revenue to the agent.

Large accounts are usually reserved for the carrier's internal sales force, and conflicts are negotiated on an individual basis. Consumer sales through agents have leveled off at 20%, because they are increasingly attracted to retail stores for portables. The revenue per month for a consumer is $50, compared to $100-120 for a business user. In addition to lower usage, there are consumer pricing plans that provide a decreased monthly access fee combined with a higher peak-period usage rate. This also accounts for part of the lower monthly average consumer billing.

PRICING AND EQUIPMENT

Leased equipment and service are often bundled in advertisements but broken out separately in billing. Because prices for telephones have come down so low, bundling is no longer a major consideration. Instead, agents discount the phone with part of the sales commission or sell it at cost when they cannot get full price. In California, the state public services commission (California Public Utilities Commission, or CPUC) has prohibited the subsidization of the telephone price with commissions by requiring that telephone prices cannot be contingent on service subscription, in response to complaints of resellers who cannot compete with carrier commission levels. Agents and retailers still discount the phone without requiring service subscription, assuming they will get 80% of the activations anyway.

Current wholesale prices for cellular telephones are about $300 for mobiles and $600-800 for the latest pocket portable models. The direct sales force of the agent will try to price the telephone about $50 over wholesale (see Figure 6–5). Various agents and dealers may offer telephones below wholesale, subsidized by part or all of their sales commissions, usually through retailers where the channel sales cost is low.

Originally, carriers engaged in equipment sourcing to obtain discounts that their agents couldn't get. However, manufacturers soon offered almost the same discounts to agents and dealers at low quantities, and carriers did a poor job of managing telephone inventories. Carriers now often buy equipment only for themselves, and merely guarantee quantities for their agents and dealers rather than handle the equipment. Carriers may contract to private-label an exclusive model with their logo for delivery only to their agents. Manufacturers

	MOBILE	PORTABLE
Price from manufacturer	$300	$600
Agent markup	50	50
Broker-agent markup	0	0
Dealer/sub-agent markup	−100	−100
Carrier major accounts	5%	5%
Carrier direct sales	5%	5%
Retailer	−100	−200

Figure 6–5 *Cellular telephone price markup*

and wholesale distributors manage inventory and service far better than carriers.

Manufacturers supply agents with product literature, demonstration models based on sales levels, and co-op at about 4% of sales. Larger agents usually buy direct from manufacturers, and buy from brokers or wholesale distributors only when an odd brand or rapid delivery is required for an out-of-stock unit. Motorola requires agents to go through distributors.

Dealers/sub-agents and retailers typically sell below cost in order to make commissions on activation. The direct sales force of agents and carriers try to make some money on the phone, but will sell below cost when necessary. The broker-agent passes the phone through to sub-agents (dealers) at cost in order to get an activation from the sub-agent. At about $300 commission, everybody makes money.

SERVICE

Customers have learned to refer usage and billing questions directly to the carrier. About 100 customer service calls per month, separate from registration and subscription, are received by the average 20-salesperson agent. Typical service calls break down as follows:

- Additional work on telephone installation   50%
- Question on new equipment operation   10%
- Dropped calls or poor reception   30%
- Billing or other   10%

Follow-up service required by customers of agents rather than the carrier is minimal and is not proportional to the residual commission. Almost all customer service rendered by the agent relates to equipment and initial installation and subscription. The carrier resolves billing and dropped call questions that have been determined to be service-related rather than equipment-related. The agent handles all equipment-related problems, registration, and feature adds and changes. If you ask a carrier or agent where customers call first for service, each will identify themselves as the customer's preferred first point of contact.

Customers expect a sales call within 72 hours and installation scheduled within five business days from purchase. A record is kept for all customers, which includes an entire sales and service history. This is used for customer support, new customer follow-up, and periodic courtesy calls for service and new customer referrals.

RECENT TRENDS

The current trend is for carriers to reduce commissions paid to agents and dealers and increasingly expand their internal channels, from a direct sales force for large accounts to a direct sales force for general business customers. Further, they may lower commissions paid for consumer customers versus business customers (based on which rate plan the customer chooses) and sidestep external channels to promote portable sales and service subscription directly to consumers.

At the same time that up-front commissions are declining, the trend is for residuals to increase to as much as 9-11% of revenue and to pay them as part of the compensation to both external and internal channels. This recognizes the value of residuals as incentives for keeping customers, and as a loyalty incentive to ensure that the agent or salesperson can never leave the carrier without great income penalty.

Carriers often pay up-front sales commissions conditionally. If a customer leaves the service before 90 days, the sales commission is taken back by crediting it against future commissions. As revenue per customer is dropping, carriers are extending these conditional periods to 120-180 days.

AGENT CONTRACTS

Agent contracts are standard for the traditional agent, which is a full-service facility dedicated to cellular with a direct sales force. The only thing that might vary is the commission level. Even though this might be set out in the contract according to volume, a carrier might pay a higher commission for the same volume to get an attractive agent. This is not recommended, as it eventually forces the company to pay that rate to everyone. Of course, the commission rates change during the life of the contract many times anyway due to changing market conditions.

A typical monthly quota for a new agent would be 200 gross new subscribers. This is based on a sales force of 10, each making 20-25 sales.

A standard agent and a broker-agent can have the same contract. Sometimes the standard contract will be modified to state how much of an agent's business must be direct versus through dealers in order to assure a high concentration of direct sales for certain agents. Broker-agents may get a lower commission because they may be involved with dealers who churn customers from one carrier to another, or otherwise have no control of the quality of the end user.

A contract with a retailer must be separate, because the carrier cannot control a retailer to do business as the carrier wants. A company like Sears or Radio Shack will dictate to the carrier its terms, rather than the other way around, and a custom contract usually has to be designed for each retailer, based on a model called an "Authorized Retailer" or "Authorized Sales Center" contract. Retailers often will not allow any provisions regarding required services to be performed by the retailer as agent, the way the retailer may be inspected or supervised by the carrier, restrictions on which cellular telephone product lines must be sold (or requiring them to carry private-label merchandise of the carrier among others), or quota requirements. The carrier is usually happy to get retailers even on these terms, because of the value of the retailer's own name and presence in the consumer segment.

AGENCY TERMS

The agency contract is virtually always exclusive with one carrier, otherwise the entity becomes a dealer under a loose contract or no contract at all. The commission terms, based on the sales quota and previous performance, have already been described. The contract usually describes the requirements of the agent in very specific terms, and assumes that the agent has been carefully accepted for its role by the carrier or reseller. These requirements include the following:

- The size and location of the agent's facility, and the operations it must perform in terms of sales, customer service, installation, equipment inventory, computer systems, and so forth.
- The agent making cellular its primary or only business at the location, and subscribing customers exclusively for one carrier or reseller.

- The carrier having approval rights over advertising and promotions and the way in which the agent may use the carrier's name and logo.
- The sales quota, quality standards, and commission structure for the agent. It includes the conditions under which failure to achieve quota and other performance deficiencies (customer complaints, sloppy operations, etc.) may lead to severance of the contract.

Commissions

Commission terms vary based on the quota and customer base of the agent, rather than the type of customer. Residuals based on customer revenue automatically regulate the quality of the customer. Recently, carriers have been trying to implement programs to lower commissions to all agents for consumer customers versus business customers, because their usage as a class is so low. In addition, carriers are denying commissions to their internal major account salespeople for sales that are not to major accounts. Those markets paying the high end of commission levels are continuing to lower them because of a decline in usage per customer.

An agent would typically need more than 4,000-5,000 customers to get a residual of more than 5%. Commissions are lower below the top 60 markets, but not too much lower, because all over the country the low prices of phones don't provide enough margin to agents to support them. Commissions would be about $300 in a city like San Francisco and $200 in smaller cities beyond the top 60.

THE IDEAL AGENT CONTRACT

An ideal agent contract would start with a high up-front commission to attract a new agent, and move quickly to a high-residual, low up-front commission structure to keep them once they started to accumulate a customer base. It would give them incentives to keep existing customers rather than provide incentives for their sub-agents to churn existing ones.

The ideal long-term contract would have commission rates of about $150-200 per activation, and residuals of 3-7% of revenue depending on the size of the customer base. The actual contract might actually start out with up-front commissions of $300 for one year with no residual, then move to the long-term model. It is important to require that customers stay on the system for 180 days for commissions to be earned.

Sales of a pricing plan geared toward consumers (low access, high peak-usage rates) generate up-front commissions 50% less than regular. Sales of a pricing plan with a 200-minute minimum per month or a one-year minimum length provide commissions 20-25% higher than regular.

The agent contract provisions should include the following:

1. Agent exclusivity.
2. Full-service facility.
3. Increasing residuals with growth in customer base.
4. Agreement to leave major accounts to the carrier.
5. Agreement to leave general business accounts to the agent.
6. Requirement for some agents to concentrate on direct sales versus sub-agents.
7. Use the carrier logo by the agent's direct sales force only, not by sub-agents of the agent.
8. The carrier's right to offer cellular phones and other offers to its existing customer base or to the general public on a direct marketing basis, even though an agent originally sold them.

The above commission rates and contract terms minimize churn in several ways. For example,

- Low up-front commissions (with good residuals) and 180-day minimum reduce the incentives for sub-agents and independent dealers to churn customers.
- Residuals provide incentives to reduce churn, but, more importantly, make it impossible for the agent to leave the carrier and give them up.

To avoid channel conflict, the carrier (or reseller) and its agents should understand up front that the carrier will handle major accounts and direct marketing to consumers and small businesses. Its direct sales force will not cover any segment but major accounts. The carrier's direct sales force will not receive commissions for sales to other than major accounts. If an agent sells a major account occasionally, the agent will receive an up-front commission but no residual and must turn the account over to the carrier.

There are no examples where a market has been able to maintain any discipline on cellular telephone pricing. If there were some contract terms that could maintain price discipline between the carrier and its agents, it would be eroded by competitors or disallowed by law.

If agent contracts were non-exclusive, residuals would lose their importance in keeping talented agents and as an incentive to keep customers on one carrier. For maximum carrier performance under non-exclusivity, residuals would be dropped, up-front commissions would rise, and carriers would try to extend the minimum customer life to pay commission from 90-180 days to one year. Carriers who have tried to keep agents exclusive while offering no residuals and constantly reducing sales commissions have been left scratching their heads or suing their agents until they offered residuals.

Monthly performance information should be shared with exclusive agents freely as peers and partners of the carrier in the retail marketplace. The only information to be kept from an agent is the performance of other agents and of the carrier and resellers. The following typical information should be regularly shared with agents, and perhaps with direct dealers:

1. Carrier activation and churn rates for the month. This allows the agent to tell if its performance paralleled the overall carrier performance for the same period.
2. Churn.
3. Customer satisfaction: sample taken once per quarter by the carrier.
4. Breakdown of sales by market segments: by company size, management level of end user, industry, and personal lifestyle of end user (income, location, automobile, etc.).
5. Usage per customer.
6. Average bill size.
7. Mix of pricing plans subscribed.
8. Mix of mobiles/transportables/portables: once per quarter.
9. Trends in slow-pay and bad debt.

AGENT FINANCIALS

Agents have no incentive to reveal their financial condition to their carrier or reseller. In many cases, agents fear that if they are financially healthy, commissions may be reduced, or support for the channel will be withheld. Relationships between agent and carrier may be strained because the carrier competes directly with its own agents, provides little marketing support, or attempts to reduce costs through lower commissions. They also feel that giving the appearance of poverty may discourage new agent entrants. At any rate, agents have few incentives to demonstrate financial health, especially where the relations with the carrier are not viewed as a partnership.

Contracts with agents should require quarterly financial statements, although even these will always appear to demonstrate that the agent is not making any money. The carrier should review these to make sure that agents are successful, rather than to look for opportunities to reduce commission costs. Only a long-term development of partnership and trust can permit the carrier and agent to function on terms of mutual support.

The illustrative agent income statement is developed on the current performance assumptions in Figure 6–6. Actual agent performance and characteristics can be used, like these, to develop a model that produces the income statement in Figure 6–7, and these amounts can be easily estimated even without the agent's full cooperation.

It is apparent that this fictional agent utilizes equipment sales as a tool to sell service, rather than as a profit center, although the CPE is not subsidized. The residual commissions cover a large portion of fixed costs, while the current sales rate supports the variable portion of operations costs. Some sensitivity tests will show that the profitability of this agency is dependent on how well it can maintain a large customer base at low fixed cost. Thus, long-term profitability is not simply dependent on sales growth, but on keeping churn low so that sales are not absorbed by churn replacement. If churn is kept under control, residual commissions are pure profit.

If the sales rate and churn rate remain constant, this agent will "top out" at about 10,000 customers. Each year the income statement will change drastically, based on the relationship of the sales rate to the total customer base and the sales commission and residual commission rates that go with each. Each agent, and all channels, must determine the optimum long-term size of the operation that provides the best balance of the carrier's or reseller's quota demands, with the overall objectives of the agent's business. For example,

1. Maximize gross profit.
2. Maximize percent return on sales.
3. Maximize percent return on investors' capital.
4. Maximum manageable operation size.
5. Maximize growth to position business for sale

AGENT FINANCIALS

ILLUSTRATIVE AGENT INCOME STATEMENT		
Assumptions/Current performance:		
Customers	5,000	
Sales rate	250	per month
Churn	0.025	2.5%/month
Phone sales	0.8	80% of activations
Sales staff	10	
Sales rate	25	per salesperson
Commissions	$225	per sale
Residuals	0.05	5% of revenue
User revenue	$120	per month
Phone margin	0.1	10%
Staff	$2,000	per month per person
Rent	$3,500	per month
Installs	0.1	10% of phones sold
Install. price	$125	
Phone price	$500	wholesale
Install. labor	$40	
Install. material	$50	
Sales comm.	$125	per sale
Marketing	$20	per sale
Management	$4,000	per month
Calculated figures:		
Salesperson compensation @25 sales/month= $37,500 per year		
Churn/month	125	

Figure 6–6 *Agent income statement assumptions*

ILLUSTRATIVE AGENT ANNUAL INCOME STATEMENT				
Revenue				
	Equipment sales			$1,320,000
	Installation			30,000
	Commissions			
		Activation		675,000
		Residual		360,000
	Total Revenue			$2,385,000
Cost of Goods Sold				
	Equipment			$1,200,000
	Installation			21,600
	Total Cost of Goods Sold			$1,221,600
Gross Margin				$1,163,400
Expenses				
	Sales personnel			
		Salaries		$0
		Commissions		375,000
	Customer service		3 people	72,000
	Administration		3 people	72,000
	Facilities			42,000
	Marketing programs			60,000
	Management		3 people	144,000
Total Expenses				$765,000
Operating Margin				$398,400
Operating Margin Percent of Revenue				16.70%

Figure 6–7 *Agent income statement*

In summary, the agent channel originated with the first commercial cellular service and is still the largest and most important channel to cellular. There are literally thousands of agents serving cellular nationally, and their numbers continue to climb rapidly. While their role is still evolving and the need for technical and installation service is eroding, there is no sign that the channel is beginning to be displaced by retail and internal channels.

CHAPTER 7

ADVERTISING AND PROMOTION

DEFINITIONS

Advertising includes any paid use of media to promote a product or company. It is usually included in the broader category of marketing communications, which also includes public relations and product and sales literature.

Promotion is an incentive or offer specifically designed to bring customers to the final stages of the purchase decision. It may be presented through advertising of a specific offer or a public relations event, or offered through the salesperson only.

PROMOTING INTANGIBLES

It is natural for the cellular carrier or reseller to introduce and promote the service through advertisements that develop the themes of the benefits of cellular service alone. But, like other intangibles, it is difficult to sell cellular as a service without the aid of tangible images of its product's characteristics and benefits.

The provider of any product or service naturally believes that its products are the primary focus of a customer purchase decision, rather than a component. Intel[1], for example, likes to believe that customers choose personal computers based on the fact that their chips are inside ("Intel Inside" on the box). Likewise, cellular carriers believe that prospective cellular customers decide to have cellular service for all its benefits and need a cellular telephone in order to use it.

However, such is usually not the case in cellular. Time and again it has been demonstrated that the customer's focus in cellular is based on the decisions regarding the features and price of the cellular telephone, and that cellular service is a necessary but secondary consideration. While prospects need to go through the steps of exposure, awareness, and so on in their decision to become a cellular user, they almost always think of the telephone instrument as the primary component and "deliverer" of the service. Their purchase decisions are based on the choice of telephone, not the service provider. The customer will most often subscribe to the cellular service provider associated with the seller of the phone.

PROMOTING THE CELLULAR PHONE

Thus, a direct sales force for a carrier, or its exclusive agent, will get the business based on selling the customer the phone. Resellers and independent dealers who are not aligned with one carrier will be able to influence the customer who buys a phone from them to subscribe to whichever carrier they recommend. This has important implications for both channel strat-

---

[1] Intel and "Intel Inside" are registered trademarks of Intel Corporation.

egy and promotional strategy. The carrier cannot control sales channels by limiting who will be allowed to promote the service. Instead, it must attract those channels who are most successful at selling telephones, the focus of customer attention.

The Paintbrush Analogy

In order to properly understand this phenomenon, I like to use the analogy of the paintbrush manufacturer. When it comes to painting, there is nothing more important, in terms of productivity or the quality of the final product, than the paintbrush. Certainly, paintbrush manufacturers are proud of the quality of the paintbrushes they produce and would like potential customers to know it. However, no matter what they might do to convince people that their paintbrushes are superior, that people should seek out stores that sell them, and that they have the best variety and price available, people never make the paintbrush the focus of their painting job decisions. People rarely even consider paintbrushes when painting except as an afterthought, unless they're a professional painter. No matter how important and rational it is to use the right brush, people will always begin a painting project by selecting a color. They may remember the paintbrush just before they walk out of the door of the paint store. No matter how irrational consumer behavior may be, it is extremely difficult to change.

In the same way, while some customers may actually make their purchase decision based on cellular service, the service provider must remember that the customer most often will come to purchase cellular service by first being brought to the decision through the vehicle of the tangible, visual cellular phone. It may be neither rational nor practical, but it is virtually impossible to change. It is easy to work with these assumptions as long as we don't let our ego get in the way and try to force the customer to be attracted to cellular by the service.

Lessons for Promotions

There are three lessons here about the promotion of cellular through the attractiveness of the telephone.

1. Customers believe they have obtained "cellular" wherever and whenever they buy a cellular phone, not just where the carrier's branded service is sold or its authorized sales outlets are located.
2. The customer associates cellular service with visual images of the cellular phone.
3. The consideration, and thus the offer, of cellular service is subordinate, secondary, and after the sale of the phone.

ADVERTISING

Advertising serves to attract people to consider cellular and to position them against competitors. However, while it is tempting to position the customer using the benefits of the cellular service, it has been found that advertisements are not successful unless they include some pictorial aspect of the cellular telephone instrument. Without the telephone, the customer cannot visualize cellular effectively in an advertisement.

Educational Advertising

The need to portray the telephone in advertising applies to both initial educational advertising, at the introduction of cellular service, and to competitive and promotional advertising, as the market begins to grow and mature. Cellular markets appear to reach a point where the base of customers is large enough to cause word-of-mouth to become a more powerful stimulator of interest in cellular than any combination of benefits advertising and public relations work. As a rule of thumb, *a market reaches a "critical mass" at approximately 0.5% penetration.* At this point, cellular has credibility as a valuable service in the market, and educational aspects of advertising cease to be a draw. Early adopters have mostly accepted the new service and are opinion leaders for their peers. As we alluded to in the chapter on market segments, potential customers learn of the benefits of cellular and are drawn to the first stages of adoption—exposure and awareness—through several sources.

1. Interaction with opinion leaders among their peer group, or business associates who are already using cellular.
2. Increasing frequency of receiving calls from people who, based on the background noise and characteristically "rough" cellular reception of newer markets, are obviously using cellular.
3. Increasing frequency of receiving calls from people who compulsively *tell* them they are calling from a cellular phone. (While sometimes this is a explanation for the quality of the reception or the need to keep the call short, often it is merely a power trip.)
4. Being a passenger in a car equipped with cellular.
5. Increasing frequency of spotting cellular's characteristic "pigtail" antenna on automobiles.

Once this stage of growth has been reached, probably in the first 18-24 months of service, customers use advertising to make a trial purchase decision, rather than merely to understand cellular and become interested in it. As far as awareness is concerned, the medium becomes the message. The customer who has not thus far become aware of cellular as a benefit may not always notice the content of the advertising, but notices how many advertisements there are and realizes that the concept is for everyone (perhaps including him or her), rather than a fad for a special few. Another benefit of encouraging multiple sales channels is the appearance of multiple advertisements from different providers in multiple media, each getting to different kinds of customers with different promotional messages.

Promotional Advertising

The content of successful advertising in this stage begins to concentrate on the trial purchase decision, focused on the telephone rather than the service. The ads provide "ease of entry" through leasing the phone rather than purchasing it and time-limited price offers, while differentiating the features of various telephones or offering free premiums with the purchase of a phone (CD player, extended warranty, speakerphone or other accessories, etc.).

The customer is brought to the decision to purchase through the avenue of a choice among attractive offers, whether conscious or not. The sales channels need such offers as one method of producing highly qualified sales leads, although leads generated solely through price offers usually attract customers who are price-sensitive and already predisposed to buy. (The direct sales force should be directed at those customers who constitute a target market with a high need for cellular, who need to be convinced, and who are not price-sensitive.)

Measurability

Another feature that recommends a time-limited, specific offer of price or free premium, is its measurability. The success or failure of the advertising can be measured easily through the sales success with prospects who were led to the sales presentation through the advertisement.

The directness of the promotional offer is an important aspect of its immediacy and effectiveness. By directness we mean how many steps are involved between response and purchase. Media advertising or a direct mailing that invites a customer to return a card or call to obtain a brochure introduces a first step. The customer then receives a mailer or brochure containing service information or a specific offer. This may invite the customer to call or return a card as a second step. Often, a follow-up call from sales removes the requirement for the prospect's initiative for the second step.

The law of diminishing returns applies severely to such programs. If the response rate for a mailing averages 2% (high), the response from two-stepping a response is at best .02 × .02, or 0.04%.

Even if a customer responds to an advertisement by requesting to subscribe or to have a salesperson call on them, a requirement to have someone call them back constitutes two-stepping and diminishing returns. Anyone handling calls in response to an advertisement should be able to initiate service or set and appointment for a sales call, rather than have someone call them back. A "live" handoff to a salesperson or customer service representative, rather than a "warm" lead through a call-back, is mandatory.

Time-Limited Call to Action

The advertising content must provide not only a direct response mechanism, but a unique code identifying the offer and the medium for tracking purposes. In addition, a time-limited offer (e.g., "Before June 30" or "For 30 days only") urges a call to action. Only a single channel should be offered for response, although it may have several agents or convenient locations within a single channel, or a single call to locate the nearest agent. Advertisements that invite customers to call the carrier, see the nearest agent, and call for the nearest dealer are not only unfocused, but difficult to measure. Printed advertisements often include a mail-in reply option. However, over 90% of replies are by telephone, and these usually reflect higher immediate interest in immediate purchase. Mail-in replies reflect "softer" interest and are usually superfluous.

Single-Minded Proposition

In the same manner, advertising copy must be limited to one offer, to one customer segment, and through one channel. It is tempting to extol the virtues of cellular service, illustrate several applications and the attractiveness of several offers, and emphasize the quality of service offered at multiple locations in one advertisement. However, the desire to overwhelm the customer with the superiority of the carrier's overall offering through multiple messages merely diffuses the importance of each one. The stimulation of interest is much more effective through multiple exposures of the potential customer to different offers, in different media, through different channels over a short time period, each promoting one benefit, one offer, at one place. This occurs partly through good management and partly through serendipity when multiple channels are used. On any day the potential customer might see one advertisement promoting the price of several cellular telephones, another promoting the features of one manufacturer's newest unit, another stressing quality, local installation, and service, and a fourth with a free premium.

In the mature market, more and more cellular users become repeat buyers, and their knowledge and experience permits them to discern the differences between carriers more carefully. But they are too difficult to target to consider promoting the service instead of the unit, and they still require the visual image of the phone. Promotions of free airtime merely cause customers to switch from one carrier or provider to another and do not attract good, new customers. Repeat buyers who need an additional phone will be attracted to the same promotions that attract first-time buyers. They will go back to where they were treated well, and shun carriers and dealers who have treated them poorly.

MEDIA

As mentioned, the core target market for cellular is the mobile small business executive. However, as individual markets and the cellular concept mature, much broader segments are attracted. To reach the core target market, media must be targeted as well as the advertising message. This not only efficiently directs the advertising to the target audience, but also controls marketing costs. Television, for example, as a mass-market medium, is not usually appropriate until the market is mature. Television is not only expensive in media costs, but also in production. This is evidenced by the low quality of local television commercials, usually "starring" the proprietor.

The mobile small business executive is appropriately addressed by radio in the automobile during commute hours ("drive time"), especially in conjunction with business news and traffic reports. Advertising in the newspaper is most effective in the business section and in the sports section, where demographics of recent new customers indicate a male audience is desired. A mix of alternating newspaper and radio advertisements directed at the same offer or benefit reinforce each other for a particular campaign and is usually most effective. Experience will determine appropriate days of the week for newspaper advertising (usually the beginning of the workweek for cellular) and the best stations and times for radio. As markets mature, the advertising and promotional message expands to include other sections of the newspaper and radio stations with music formats preferred by the target segments.

Television has been adopted in some cases by carriers in mature, major markets where consumer segments are being significantly penetrated. The effectiveness of television must be measured even more carefully than other media. With television advertising directed at consumers, we encounter a phenomenon where the most expensive advertising medium is being used to attract the least valuable customer in terms of monthly revenue per user, customer longevity, and customer loyalty.

PUBLIC RELATIONS

Cellular has always been able to capture media attention through its novel applications, high-tech interest, and public interest angles. Cellular has increased highway safety, traffic reporting accuracy, and crime prevention to name a few public benefits. A good public relations person can supplement advertising and word-of-mouth public awareness with frequent news releases concerning new technology, coverage additions, cellular use by public agencies, and so on.

The carrier must recognize the role of cellular to improve public services and the public good and contribute to interests such as emergency communications plans and other public needs that cellular can serve well. While the carrier's responsibility to promote the community interest should be at the forefront of the carrier's participation, the resulting good publicity can be an excellent supplement to the marketing program.

ADVERTISING PITFALLS

There are several pitfalls to avoid when advertising, some of which apply uniquely to cellular.

1. The "burning eyes" syndrome: It is easy to fall into the trap of advertising heavily in response to the highly visible advertising of competitors, national cellular equipment manufacturers, or independent sales channels. "Burning eyes" comes from the way you or your boss react when you see such advertising, which may demean your own offering, make inflated claims about your competition, or copy your strategy. Resources may be thrown at creating a counter-campaign that is equally visible and expensive as the one that stirred your angry emotional response. This can cause a cellular provider to throw its entire annual advertising and promotion strategy and budget out of kilter to no purpose. Advertising must be guided strictly by the objectives, schedules, and strategies in the annual marketing plan.
2. Advertising measurements: The measurement of advertising recall, brand awareness, and other aspects of the advertising program are normally effective in the market for retail consumer goods. In this market there is no

direct measurement of the effectiveness of an advertisement in producing a sale, because the sales channels are not uniquely associated with the products. If the advertising is correctly aimed at sales generation, such measurements are subordinate and unnecessary, compared with direct measurements of inquiries in response to specific offers.

3. The desire for visibility: All advertising attempts to "move out of the clutter" with larger, more visible newspaper advertising; more frequent, creative, and varied radio advertising; and the visual and leadership appeal of television. By measuring effectiveness rather than serving these desires, the advertising program will better achieve its goals.

4. Over-management: Everyone in the company will want to approve advertising and recommend advertising strategies. In addition, everyone in the company will criticize advertising based on how it appeals to their personal tastes. It is important to remember to be objective and concentrate on the tastes and response of the target market. Most important is for senior executives to stay away from personal input on advertising, insulate marketing from criticism from other departments, and yet be merciless in the demand for improvements in objective measurements of results.

AUTO-MARKET RESEARCH

Finally, all marketing managers and corporate executives are cautioned to carefully discipline themselves to refrain from what I call "Auto-Market Research," the projection of personal preferences on the market as a whole. This is a condition where managers assume that their own tastes are congruent with those of the target market and substitute such opinions for objective research. This results in acceptance of new advertising programs and new telephones from manufacturers, for example, without a proper evaluation by potential customers.

Always rely on market research and related focus group reaction to advertising, promotions, service features, impressions of customer service, sales presentation quality, and all parts of the marketing process. Never ask your boss for an evaluation of something on which the targeted audience is the better judge.

CHAPTER 8

MARKETING MEASUREMENT

THE DISCIPLINES OF MEASUREMENT

As with any good marketing program, one of the most important aspects of cellular marketing is the measurement of marketing efficiency and effectiveness. While most will admit its importance, the implementation of a disciplined program to measure marketing effectiveness and also act on it is rare.

Congruents

There are many components of marketing effectiveness to measure, which include the following:

1. Revenue versus sales expense.
2. Customer gain versus sales expense.

3. Revenue versus marketing expense.
4. Customer gain versus marketing expense.
5. Lead generation versus advertising expense.
6. Customer satisfaction.
7. Customer churn.

This list is not exhaustive, and many more elements and combinations might be included. But it will serve to illustrate the meaning of various measurements and how they are applied to decisions and marketing actions. Let's look at each more closely.

1. Revenue versus sales expense: This is an important measure but has little to do with marketing effectiveness in a service environment. In a manufacturing environment, revenue is achieved through sales booked during the current period. Thus, there is a direct relationship between current sales expense and current revenue. In a subscriber service, however, the sales effort results in future revenue over the life of the customer (the length of time the customer subscribes). Therefore, the measurement of sales expense versus revenue is an important diagnostic of current financial performance, but does not measure sales performance.
2. Customer gain versus sales expense: This is an important measure, but usually is measured as *net* gain (new adds minus churn) instead of *gross* gain (the actual number of new customers sold during the period). It is measured as net gain because, from the company perspective, net growth is the bottom line for company growth and financial gain. Also, it prevents sales channels from inflating results by activating "imaginary" customers who are dropped after a short time. Unfortunately, the sales channels have virtually no control over churn.
3. Revenue versus marketing expense and
4. Customer gain versus marketing expense:
These are analogous to the first two items, but for marketing instead of sales. While it is good to measure market-

ing and sales separately, there also should be a measure of them combined. The two areas may be organizationally separated, but they are really a single function. Sales is one portion of the marketing mix.

5. Lead generation versus advertising expense: This measure recognizes that advertising must be directly measured for results. Optimally, the actual sales that result should also be measured. While much of the sales rate may be outside of the control of advertising pull, the advertising may be pulling customers who are in the wrong segment or unusually price-sensitive, for example. Thus, even a promotion that provides excellent lead generation may not result in closed sales comparable to other promotions that provide fewer or lower-quality leads, but better sales results.

6. Customer satisfaction: See the section on Customer Satisfaction further on.

7. Customer churn: Churn is an excellent example of a measure that everyone watches and worries about, but about which little can be done. Most of the time, it is thrown on the shoulders of sales, as mentioned above, to make gross gain net gain, even though sales has no influence on churn. Sometimes, it is used as a measure of customer service quality, although no one has been able to show that churn is substantially reduced with better customer care. Unless churn is over 2.5% per month, probably little can be done to improve it that does not show up in customer satisfaction measurements. If it is under 2.5%, chances are the carrier is not actively pursuing new market segments aggressively. This may be a sign of marketing and sales complacency, rather than a positive measure of control.

Providers with low churn will say that their customer care program is at the root of it, but that is more public relations than cause and effect. While various measures should be taken to control churn where it is caused by competitive pressure or customer dissatisfaction, providers should understand that, in general, only the churn above 2.5-3.0% per month can be controlled. As providers continue to push subscription by recruiting market segments with marginal need for cellular, churn will rise.

Basic Rules of Management

There are three basic rules for measuring results in cellular marketing. They apply to both market research and marketing and sales performance.

1. Follow Steuernagel's rule: "What would you do if you knew?"

    Before you measure any parameter, think of the wildest range of values it could possibly assume and what action the company would take in each case. If you can't imagine the company taking action, or management arbitrarily does what it pleases, don't spend the time and money to measure it. For example, if overall customer satisfaction dropped from 90% to 80% (the absolute bottom), would you do anything to correct it? Would your research tell you what area you need to correct?

2. Measure only the parameters that are controllable. Measure organizations based on the things they can control.

    Gross customer gain is controllable by sales and marketing; net gain is not. Sales leads generated by advertising are controllable by marketing, but closed sales resulting from advertising are controllable only by the combined marketing/sales team.

3. Avoid the temptation to measure things because they are measurable.

    Revenue per customer is measurable but not really controllable (unless you restrict marketing to high-usage segments, in which case total revenue is reduced). On the other hand, total revenue is measurable and controllable.

    Sales cost as a percent of current revenue is measurable, and useful for finance to measure this month's cost performance, but not useful to measure sales performance. No amount of sales effort will significantly raise this month's revenue in the cellular environment. Cutting sales cost will lower it as a percent of revenue for the month, but this measure will rise severely due to reduced revenue in the future if sales effort is lowered today.

Relative Channel Performance

Carriers often use measures of how well the sales force is managed, rather than whether the right channel is going after the right target. The carrier measures sales per salesperson, cost per sale, and sales per agent. Rather than compare one channel with the other, it tries to improve each number from one month to the next by managing the measure and the people. Agents use similar measures rather than aggregate measures, using the statistics of individual salespeople and sub-agents to brow-beat them into doing better.

Carriers often don't want to know how high their direct sales channel cost is, when compared with other channels, because they want to control sales and don't want any reasons not to.

USEFUL MEASURES

There are two basic areas for understanding the measurement of marketing success. *Effectiveness* is the degree to which the marketing process (advertising, sales channels, everything) is successful in reaching all of the targeted potential market and converting them to customers at a rate consistent with projections. *Efficiency* is the measure of long-term revenue versus the total cost to get and keep the customer—the ratio of unit output to input. Often, marketing is responsible for effectiveness while staying reasonably within budget. Without diminishing the importance of effectiveness, efficiency is the source of improvements in effectiveness and profitability, because it frees dollars for the resources to be more effective.

An organization that gets one new cellular customer at a cost of under $100 is remarkably efficient, but not very effective. An organization that exceeds customer gain projections by 20% is very effective, but not very efficient if each new customer carries a marketing/sales cost of $500 or more and a low monthly usage rate.

There really is no substitute for revenue and revenue growth, in relation to cost, as an overall marketing measure. All other measures are subordinate indicators of where the problems are. Some measures that are useful, controllable, and "actionable" include the following:

- Gross sales per salesperson (versus peers in the same channel)
- Gross customer gain
- Increase in customer gain
- Increase in revenue
- Leads and sales generated by promotional campaigns (versus previous similar campaigns and versus expectations)
- Revenue per customer over the average life of a customer in each market segment and channel (Expected Revenue per Customer)
- Expected Revenue per Customer / Sales and marketing cost per new customer

GENERAL MEASURES OF MARKETING SUCCESS

The cellular provider must have a few measurements of immediate performance that are available daily or weekly. There are really only two categories: customers and usage. Usage is more important than customers; however, not much can be done about it. (I have always tried to correlate periods of high cellular usage with news items and other events to try to understand what causes it and what can be done to keep usage high. I have found that there are only two items I need to create to ensure high cellular usage: traffic accidents and bad weather.)

It is possible to increase usage with new applications like voice messaging and information services. This would be unnecessary if only high-need, high-usage customers were attracted to cellular. Concentrating on high-usage customers gives more immediate results and leads faster to higher revenue. Unfortunately, too often we are given incentives to increase customer counts indiscriminately for the benefit of the investment analysts and our superiors, and then asked to apply ourselves to the tedious task of making users out of them.

The most basic measures of marketing and sales for diagnostic purposes are daily reports of customer gain and usage levels. Results for the previous day have to be available the first thing in the morning so that appropriate adjustments can be made to advertising, lead generation efforts, sales force levels, commissions, and so forth. These tactical adjustments are important but cannot take us away from the longer-term tasks of intelligent planning, measurement, and strategic decisions.

Remember that having instant information is only important if you have the capability to make instant decisions and take quick action. Gather only the information that enables the tactical decisions the recipient is allowed to make, in the time frame the organization permits.

It would appear that carriers generally pay far too much attention to customer gain, especially as revenue per customer decreases. Perhaps the only measure more useless as a gross measure is "pops," or potential market as expressed by market population, used for investments in cellular markets. Direct attention to revenue gain through additional services, usage, and price increases, in addition to customer gain, are all important. Too much emphasis on customer gain will force marketing to attract more marginal customers with low usage at higher sales costs per customer.

MEASURING PROMOTION SUCCESS

The primary, short-term need during a promotional campaign is to measure the effectiveness of a particular advertising/promotional campaign by monitoring immediate response on an hour-by-hour or day-by-day basis. The promotion must be designed to have a specific order form, telephone response number, customer service team, mail order box number, or whatever means it takes to identify responses to particular programs. Additional information, such as time of day or person responding, may be appropriate, depending on the ability of the organization to take tactical advantage of more detailed information or to move more quickly. If multiple media are used, the medium must be identified if other information is the same.

Information Gathering

The identification of the demographics of the customer are just as important as the information collected in order to complete the sale. However, the collection of information from the customer must never interfere with the sale. You can probably only find out three or four characteristics of the customer before the questioning becomes intrusive. Make sure you don't have to ask where they learned about the offer. The means of reply should tell you that.

The importance of tracking a promotion is not just to measure its success. It also collects prospects for future promotions if they do not buy, for example. But direct marketing is almost synonymous with database marketing. It is the use of a database to collect not only sales information, but information on the promotion whether it is successful or not as feedback into an ever-improving cycle, that makes it an "iterative enhancement" of the direct marketing process. Even if the promotion is successful, careful measurement of the results will tell you whether or not the promotion hit the target market. If it missed but was still successful, then you have hit another market, and the original target is still unpenetrated.

If the promotion was not successful, the data collected should tell you whether the target was missed or the offer was wrong. Either way, each successive promotion should be more productive, based on careful tracking of the responses of different target markets to different types of promotions.

While the direct marketing (telemarketing, direct mail, catalog, database) aspects of promotion are important for both lead generation and closing, instruction on all aspects of these topics is beyond our scope. Good books on direct marketing will help with technique; our purpose here is merely to recommend that some techniques of direct marketing and database marketing are appropriate for measuring and improving promotions.

CUSTOMER SATISFACTION

Customer satisfaction studies are important as a measure of quality and the various components of customer care that, if neglected, can lead to increased churn, low usage, and lack of adoption of the cellular concept over the first few months as a cellular subscriber. Such questionnaires can also include questions about demographics, which should verify data obtained by other means about the preferences and proportions of customers in each market segment.

Customer satisfaction studies are also very easily manipulated and should always be conducted by an outside market research firm, as with all market research. When properly conducted, customer satisfaction can be used as a performance measure if it can be correlated with long-term improvements in revenue per customer and customer loyalty. It is also very easy to produce a customer satisfaction study in which 90% or more of the customers rate overall service as satisfactory or better. What is really important are the sources of dissatisfaction for the remaining 10%, and statistical confidence that satisfaction is improving.

Detail Requirements

In order to provoke action, questionnaires must break down the sources of customer satisfaction and dissatisfaction, so that specific sources of dissatisfaction can be corrected. Typical questions for the customer satisfaction study would include the following (rated on a 4-5 point scale of excellent, good, average, below average, poor or something similar):

- Overall satisfaction
- Ability to access system
- Billing detail and clarity
- Size of service area (and areas of poor service)
- Dropped or unsuccessful calls over the past 30 days
- Interactions with customer service staff, if any
- Cellular service as a good value for the money
- Open-ended comments

Other questions would include experience with salespeople and technical people. Too often, the content may get bogged down in questions about whether the salesperson gave the customer instruction or information in particular areas. This often occurs as an overreaction to learning of some customers being misinformed about prices or operation, for example. This results in the company requiring salespeople to inform customers to distraction about things they don't want to know. Asking a customer whether he or she was informed about roaming procedures does not tell you whether they wanted to hear about it. It is bad enough to make a customer a victim of your attempts to discipline salespeople, don't make them the scorekeeper also.

Longitudinal Studies

A customer satisfaction study, as shown by the sample questions above, applies to other areas besides sales (coverage, service quality, etc.). The study should be conducted on a regular basis, using the same questions and scoring, so that changes can be fairly represented. The market research organization can show you how to conduct a "longitudinal" study, in which the same customers are checked for changes in attitude over time. This should be considered in addition to ongoing studies of different, random samples, and special studies of important subgroups of customers.

SUMMARY

As with other areas of measurement, here are two reminders:

1. Don't study anything you're not willing to change. Examine the range of "What would you do if you knew?" before embarking on a study.
2. All marketing, sales, and customer satisfaction measurements are subordinate to financial results. It must be shown that any improvements in marketing measurements will lead to improved financials.

Finally, remember the financial responsibilities of marketing to keep costs down and revenue up. Ratios of life-cycle revenue to sales cost are important to determine relative efficiency, but raw costs per activation, like the typical ones in Figure 8–1, are much more urgent in their message that costs are too high in a business where revenue per customer is eroding.

	AGENT	CARRIER
Sales cost	$250	$350
Loaded	$400	$550

Figure 8–1 *Cost per activation: Retail*

CHAPTER 9

COMPETITIVE BEHAVIOR AND ANALYSIS

Many easy analogies of marketing competition are made which liken competition to war, to a battle ground with fight-to-the-death combat, to winning and losing. These are important aspects of competition. The portion of competitive thinking that assumes that every sale the competitor makes is one that you could have had is an important part of aggressiveness and attention, which competitive vigilance requires.

At the sales level, competition is most often a head-to-head bid for the same business from the same customer. There are several aspects of competitive behavior, however, which good marketing techniques can contribute to make competitive superiority easier and less risky.

COMPETITIVE DIFFERENTIATION

The competitive threat is worst where all competitors offer the same product at the same price to the same customers. It is important to differentiate cellular service from the offerings of all others in the marketplace. But even differentiation means little for marketing purposes unless the customer can see it. This is done in the following ways, for example:

1. Product definition: The product can be distinguished and differentiated by the features it includes, even though they may be intangible—the size of the service area, inclusion of vertical features or voice messaging, the refund policy on dropped calls, the information services like traffic and financial reports, etc.

2. Promotion: The product feature doesn't exist as a distinguishing factor unless it is promoted heavily. Make sure you are willing to concentrate on these factors in advertising (other than CPE promotions), literature, public relations displays, and sales presentations. The value of a distinguishing factor as a promotional item can far outweigh its real importance to users. Fiber optic trunks, for example, don't by themselves mean a thing to user benefits, but are an excellent distinguishing product feature to publicize if the other carrier doesn't employ them, as a perceptual indicator of the carrier's dedication to quality and latest technology. Sprint, of course, has done this with its long-distance network.

    In addition to distinctive messages, promotions must have a distinctive look that becomes the trademark of the provider as much as its logo. The announcer's voice, a familiar spokesperson, the size and placement of advertisements in newspapers, colors, and types of pictures and fonts, all contribute to a certain look. This does not mean that whatever you do first must stay. When you find a good one, you'll know by the compliments. Then stick with it for as long as it works.

3. Pricing: Pricing is the most visible distinctive factor of a provider. By having a rational set of rate plans with

meaningful names that relate to customer groups and benefits, which can be easily distinguished, a lot of competitive advantage can be gained where the competitor is not as astute in pricing.

4. Distribution: The quality of your salespeople, internal and external, is important, but the difference should also be identified with unique symbols: colored blazers, large logos, or a distinctive color scheme for all agent locations, for example, tie together the quality with common visible reminders, and show how ubiquitous and uniform the team is.

5. Branding: Most carriers have missed the point of branding. Wirelines use their company name because they believe their telephone company name is important, and it is. But carriers, wireline and non-wireline alike, have missed the opportunity of branding their retail service separately from their carrier service. Just like Maytag Corporation sells washing machines at wholesale generically to Sears as an OEM supplier, and under their own brand name through different outlets, carriers can use the corporate name as the generic name of the service to make the most of the corporate reputation, and a separate name for premium retail services.

While many carriers have given brand names to their information services and voice messaging services, they have failed to develop a brand strategy that distinguishes all the benefits of their retail offering from the carrier offering. The retail service differs in the quality of its sales representatives, customer service, billing, and pricing packages at the least, and the reseller often doesn't want to refer to the carrier's name, no matter how much equity it has in the market.

Even a nationally franchised, trademarked name, such as Cellular One, can be used to distinguish all of these other aspects of premium retail service from the carrier name. In some cases, the name of a reseller, such as Motorola, can be stronger than the carrier's own company name, and the Cellular One name might be better used to brand retail aspects of the service. Creation of a separate name increases the apparent number of providers in the market. However, many carriers don't even recognize that there is a retail offering to distinguish from the wholesale carrier offering shared with resellers. When wireline carriers come up with a nationally franchised name, perhaps they will know what to do with it.

BOUTIQUE VERSUS INDUSTRY

Competition is not necessarily a war on all fronts. The existence of competitors in a marketplace gives the product credibility as one which is of interest to the mass market, rather than a specialty item for a narrow set of users. It provides broader distribution, multiple offering choices, local service, and competitive prices on the product and its associated accessories to more market segments. A cellular market in which two carriers "own" cellular by maintaining their own internal sales as the primary channels may be too careful at keeping competition out. In such a market, cellular appears to be a niche, or specialty item intended for a very narrow market, available in only two ways, and only in a few locations that only those who seek them out can find.

By encouraging multiple channels to participate in the market, cellular becomes a ubiquitous presence: in multiple newspaper advertisements every day, in window stickers at appliance stores, in a public relations event held by a reseller, in an open house held by an agent for dealers. Instead of two carriers battling for first place in a duopoly, there may be dozens of dealers, resellers, and agents who beat each other up every day—and love it—and keep sales costs and CPE prices low without the carriers receiving the scars.

To the customer, the existence of multiple end-user providers, some offering poor quality and low prices and others a different mix, gives the appearance of a healthy industry and a product with mass-market appeal, rather than a niche product for someone else.

Thus, while cellular carriers must compete aggressively at the retail level with their own sales forces, they must support a competitive environment that enables dealers, resellers, and other external sales channels to be successful, in order to allow the fastest and largest long-term penetration of cellular into the marketplace. In short, *carriers must work to make cellular an industry rather than a duopoly within each local market.*

COMPETITIVE ANALYSIS

The first step in being competitive is to understand your competitor. Competitive analysis is an ongoing function, especially in tactical adjustments to the daily sales and promotional activities of the competition. It is also, however, an important component of the annual and strategic plans, and must address strategic differences and current status of the company versus the competition in all areas. Competitive analysis is not an event for managers to use to show their supervisors how hard they're trying to fight competition, rather it is an honest evaluation of the facts. What are the feature advantages of their system? How well are they controlling call quality and system capacity? How is their distribution organized, in size, quality, and location? How good is their advertising agency? What is their marketing and advertising budget? What happens when you call customer service? What is the revenue and market share of the other carrier and all players in the market? What segments are they targeting?

The answers to questions such as these must be laid out side by side with your own operations, to determine where each competitor is the strongest. This analysis is the prime determinant of your positioning strategy based on strengths that are difficult for the competitor to match, plus strategies to meet competitive strengths and strategies head-on, drown them with your own unique strategy and move the competitive battle to a different focus, or sidestep them by concentrating in a different segment or marketing element.

UNIQUE CUSTOMER PROPOSITION

While we have outlined the advantages of time-limited offerings to new customers based on telephone-related promotions, the overall advantages of the company and the carrier over the competition must be portrayed to all local industry players in addition to prospective and existing customers. Rather than use periodic advertisements featuring various aspects of the service, the carrier should adopt a unique proposition that differentiates the service continually in all corporate communications, and with which customers can easily identify as a valuable, real, and distinguishing benefit of one carrier over another.

In the computer industry, for example, Tandem has positioned itself as the leader in applications requiring low downtime, adopting its name from providing computer power in tandem for reliability, and adopting its product line name and overall theme as NonStop[1] computing, which has survived years of competition as a virtual unassailable position in the market.

This unique proposition should be tagged to all announcements, advertisements, billing, and literature in the form of a short motto or statement attached to the company or service name. It refers to a feature that relies on a competitive strength of the company, and cannot easily be duplicated or compromised. Statements as to leadership, size, and length of service do not connote a benefit (e.g., "Alpha Cellular...the leader in cellular service in the Metro area"). Rather, statements should be positioned based on long-term advantages and benefits, as the examples below demonstrate.

Where coverage is superior:
   "Alpha Cellular . . . Wide-area cellular service"

---

[1] Tandem and NonStop are trademarks of Tandem Computers Incorporated.

Where the carrier is leapfrogging the competition in new system deployment:

> "Alpha Cellular . . . The new generation of cellular technology"

Where the carrier has fewer service interruptions or dropped calls:

> "Alpha Cellular . . . Guaranteed non-stop cellular service"

Where the carrier is dedicated to staying ahead of demand in system capacity:

> "Alpha Cellular . . . The Instant Access Company"

Note that these examples are benefits that are not specific to the internal sales capability or the retail customer service operation, but apply to broader benefits of the carrier operation.

COMPETING ON YOUR OWN TERMS

The carrier has to be careful to avoid competing on the competitor's terms. This can occur very easily as a result of a successful promotion by the competitor, or one that is very visible in the marketplace (see the "burning eyes" syndrome under Advertising Pitfalls). A program or communications strategy that is similar to the competition, defensive of the company's own attributes in the same area ("We have wide coverage too!"), or otherwise keying off the competition, automatically positions you as a follower or secondary player in the marketplace. Instead, you need to find a successful way to communicate a unique program that keys off your own strengths in a completely different subject area and appeals to just as broad an audience.

In a broader context, then, there is no need to compete head-on with the competitor on every attribute, or for every customer. It is far less expensive, and far more beneficial, for both carriers and their customers, to sidestep the competition as much as meet it head-on. The first step in accomplishing this is a complete competitive analysis, as described previously.

A competitor who has technical, capacity, or coverage superiority, for example, is not met by trying to increase your capabilities. Rather, your own superiority in advertising or control of quality distribution channels may far outweigh such advantages. While you need to make reasonable advances in the competitor's area of strength, it is a mistake to try to compete on the attribute for which they have demonstrated superiority.

WHO IS THE COMPETITOR?

In the early days of cellular, competition was considered in the context of competitive solutions to cellular, such as the public telephone, paging, and other mobile services. Once cellular achieved credibility as the broadest solution to mobile telecommunications access and competing systems were in place, the context of competition moved to local competition between carriers.

The development of cellular dealers as the primary retail channel has diluted the focus of competition. Carriers often consider the relationship between internal and external channels, or the promotion of telephone sales versus service subscription as the focus of the customer purchase decision, to be competitive battlegrounds. Such considerations are misplaced, and the carriers need to concentrate on competition only to the extent that it interferes with revenue growth objectives by displacing its revenue and customers. To the extent that the actions of external channels and the other carrier stimulate market growth, the other carrier may not even be a true competitor.

In the future, new digital technologies may expand capacity and therefore the growth expectations of each carrier by such a degree as to make competition for customers a subordinate concern to the need to create new applications and new market segments. With the competitive threat from new technologies and service offerings like PCS, wireless data, and digital SMR (Specialized Mobile Radio), carriers may realign competition at the level of cellular versus other mobile communications once again, rather than carrier versus carrier.

COMPETITIVE BACKLASH

Competitive response is an important threat at all times. An important part of any campaign is to think through what your competitor's reaction will be. The more successful the campaign, the more likely the competitor will react—not just to avoid losses, but because they need to save face.

While imitation is one of the sincerest forms of flattery, the best way to avoid competitive backlash is to design product features, promotions, and the like, that are difficult to duplicate. Thus, a promotion to offer a portable cellular phone with attractive lease terms might be very successful, but also very easy to duplicate. If it is a unique lease, however, offered for the first time through a unique program by a certain CPE manufacturer in partnership with a particular leasing firm, you only need to make sure it is not offered to the competitor in your market by your leasing and manufacturing partners.

I have always been amazed, for example, that non-wirelines have done so little to take advantage of their freedom from regulation over their wireline competitors in each market. While Bell entities were not permitted to offer information services, any NWL competitor could offer them and truly differentiate their company. Today, this opportunity is lost because Bell entities no longer have this restriction.

In the same manner, however, Bell-affiliated carriers are not permitted to sell, give away, or otherwise participate in long-distance services. In addition, there are Equal Access rules that require them to allow each cellular subscriber to individually choose their primary long-distance carrier. NWLs have imitated this choice.

An NWL competitor might simply bundle some (or all) long-distance service into a rate plan, and have a service which not only would provide great benefits to customers in long distance savings and billing simplification, but which their wireline competitor would be powerless to duplicate. As a unique promotional proposition, it would be unsurpassed. But the window for such a blockbuster offering may soon be closing, either by removal of such restrictions, or by imposing them on non-wirelines—or perhaps independent resellers will be first.

FLATTERY

If a an advertising campaign is unsuccessful, not only will prospective customers be silent, but so will your sales channels, associates, and personal friends. On the other hand, successful campaigns cause not only great amounts of praise from these areas, but vociferous complaints from competitors. It may be the best indication of the success of the campaign that your competitor is worried enough to threaten you. Don't be intimidated. *A competitive complaint about the aggressiveness of your advertising is the best compliment your competitor can pay you.*

DISPARAGEMENT

Don't demean the competition. It makes cellular look bad for both of you. It doesn't matter if your competitor has a lot more dropped calls than you do or if its retail channels don't have your depth of technical support and service; don't advertise it. Your competitors will merely attack your small coverage area, or some other weak point of yours. Soon prospective customers will know every possible reason in the world not to have cellular regardless of the carrier. Advertising should always portray the positive benefits of cellular and the advantages of your own offering.

SECURITY

Many companies, large and small, do not put some of their most important strategies in writing, because they are concerned about it falling into the hands of competitors (or regulators, the news media, or the general public). Business ethics dictate, first of all, that there should be nothing in your strategies that would not reflect well on the company if they were ever made public. Beyond the ethical question, it's appropriate to be very concerned about security, but it's absurd to avoid explicit, written plans about every detail of your operations because you are afraid your competitor might get them. *It is much more important that your employees know your plans than to keep them from your competitors.* Other than customer contact information and details of upcoming promotions, cellular operators and companies in general tend to overvalue information about themselves to the competition, and undervalue its use to their own employees.

OVERVALUING COMPETITIVE INFORMATION

If competitors are seriously trying to get your planning information, this is the second most sincere compliment a competitor can pay you. You are indeed a threat! It is one thing for your competitor to obtain competitive information, but it is quite another to understand its context and to act to defend against it. If you knew your competitor was developing a campaign to give away CPE, would you develop one to do the same? Probably not, unless you were already thinking along those lines (But based on what we've learned so far, we already know that such an idea is not recommended anyway, because it does not distinguish between good and bad customers, and such a program is only worthwhile to get strong customers.) Therefore, even if a competitor received confidential information, chances are slight that they would act pre-emptively on it.

HOW FAST MUST A SYSTEM GROW?

Many larger systems are constrained by capacity, yet they are scared to death of competition and refuse to raise prices. Early in system growth, you should be able to tell if cellular will grow so fast in your market that you will be unable to keep up with demand. If this is true, then whichever of the two systems is least congested will fill faster until service availability and customer bases are the same. If there is really a discernable difference in quality between the two systems, the better will fill faster until it becomes congested. The only way to match demand to service quality in such a case is to raise prices.

Whether demand is strong or not, the sales and promotion functions should be balanced to sustain customer growth at a rate that the system can support, and which does not require enormous sales and marketing resources to obtain. One look at the expected life-cycle revenue and profit per customer should tell you how much cost you want to incur to get it.

One strategy for sales aggressiveness, for example, would be to conscript every live body in the market as an exclusive dealer before your competitor does. However, this strategy leaves much to be desired. It puts too many resources after too few sales, which may eventually make the business look sour to everyone in it. It does not match sales rates to system growth and sales cost limits; it does not discriminate between professional selling and practically bribing poor-candidate customers to subscribe.

A well-constructed strategic and annual plan will make sales channels, sales costs, and sales rates congruent. Each market segment provides a different amount of revenue per customer, indicating, along with customer channel preferences, which sales methods you can afford to use for each segment.

And many good external channels must be established from scratch. You can't expect an oversupply of good candidates for cellular agents and internal salespeople in a new market; many of these have to be developed. Assuming a fixed number of quality external channel participants are available or feasible to nurture and grow, such channels may not develop for years, and the slack in the growth rate required must be made up by internal channels. If you lose the lead in developing sales channels, it is very difficult to catch up.

Those carriers which early identify the best sales channel candidates and help them succeed will have an unstoppable, long-term competitive advantage in sales.

CHAPTER 10

THE SALES PROCESS

GENERAL CONSIDERATIONS

The discussion of the sales process presumes the training of the sales force in sales planning and presentation. All salespeople should know how to qualify prospects, maximize customer "face" time, overcome objections, and be prepared.

The most important aspect of the sales process is that it must be managed, from lead generation through customer installation, as a well-oiled machine. Diligent measurement is the prime ingredient of success. Salespeople must constantly be supplied with qualified leads from outbound telemarketing efforts and inbound advertising responses. Leads must be tracked as to source, assigned salesperson, time elapsed to appointment, consummation of the presentation, time to installation or fulfillment, and final disposition of the lead. Within telemarketing specifically, measurement and efficiency are even more specific and more carefully controlled than in any other function.

With appropriate measurement, action can be taken to discover what kinds of promotions generate the most sales, not just the most leads. The most effective people in telemarketing, sales, and customer service can be rewarded and motivated appropriately. The relative value of telemarketing, advertising (media selection, radio format, time of day), and referrals from existing customers can be weighed and used to direct future efforts. The value of callbacks to prospects who do not buy can be determined, and the percentage of walk-ins resulting from various promotions can be recorded.

Each of these factors can be examined in light of their success with different market segments: consumers versus business prospects, male versus female, and vertical industry classifications.

Coordination

The success of the sales process is a function of management's ability to have telemarketing, advertising, and sales management function together to maximize their combined effectiveness. In addition, all of the functions must be working toward the broad goals of total sales development, rather than the specific goals of advertising awareness (advertising), lead generation (telemarketing), and sales per salesperson (sales management).

This is far more difficult than it sounds. Each person in each function must be measured on their individual performance within their discipline, yet each must work cooperatively toward the larger goal. And anyone experienced in the management of the execution of the process can point to telemarketing and advertising's statements of the deficiencies of the sales force in completing the sale, countered by sales' complaints of poor leads provided.

Sales Support

Sales must be supported so that salespeople spend the majority of their time in sales planning and the sales presentation. Yet in every organization, salespeople spend an incredible amount of time resetting appointments, completing forms, going to meetings, and tending to administrative matters.

Lead Generation

The sales function should spend a minimum amount of time cold-calling unqualified leads. However, lead generation is spotty in many sales organizations. In addition, support of external sales channels in lead generation is often poor in both numbers and quality. External channels do not have the ability to hire specialists in telemarketing and advertising, nor the economies of scale that the carrier has in producing a quality advertising or telemarketing program. The best operations leave promotional design to a well-run carrier organization, which supplies an adequate quantity of qualified leads to all retail channels and has the resources to put in place a well-designed and well-executed measurement system to make it work.

Campaigns

Another weakness in the direct sales chain is to forget to ensure, before the launch of a telemarketing or advertising campaign, that all involved channels, customer service, and salespeople are briefed on the nature of the campaign, the offer, and the details about when it will appear, where to get literature and equipment, how to complete the order form, measurement statistics to record, on so forth. Because of its multi-functional nature, it is common to fail to complete all of the internal communications steps before a major campaign begins.

DIRECT SALES

The direct sales process has been the leading form of sales in cellular since its beginning. With increasing awareness and acceptance of cellular, the productivity of direct sales has increased from 12-15 per month to 25 or more.

The direct salesperson's activities can be divided into sales planning and sales presentation. Sales planning involves the qualification and setting of the appointment of the prospect, and sales presentation involves the actual customer presentation.

Sales Planning

Depending on how much support in qualified leads the salesperson receives, some portion of the sales planning process includes finding new prospects. Hopefully, most of the work in sales prospecting for the salesperson involves checking back with recent new customers for referrals to qualify. Otherwise, the salespeople become telemarketers themselves most of the time.

The sales planning process can begin by reconfirming the appointment or setting it, and using the opportunity to find out more about the prospect. Some of the most important questions involve whether they are responding to a particular offer, their knowledge of cellular and interest in specific telephones, and their business or personal applications.

For each sales visit, the salesperson should develop a strategy, which includes the type of telephone to demonstrate, an order form already filled in except for optional items and incomplete customer data, the benefits and applications to discuss, and questions to ask that can help the prospect understand the benefits in their personal situation. These items should be completed for each prospect as much as possible before the first sales call of the day, so that the entire agenda for the day, and much of the next, is pre-planned.

Sales Presentation

1. Training: While most of the aspects of direct sales should be covered through an extensive training program of at least five business days, there are some aspects that should be emphasized.

    The primary reason for failing to close a sale with an interested prospect is incomplete or conflicting information about cellular. The salesperson should have complete information about the service, telephone equipment, prices, and competition. Probably the second most common reason for failure is not asking for the business from a seemingly undecided prospect.

2. Professionalism: Certain aspects of the sales presentation should go without saying: promptness, courtesy, and appropriate dress. The presentation should begin with a conversation regarding the prospect's lifestyle and business habits, and an evaluation of his or her understanding of cellular use and pricing.

3. Benefits: Benefits should be discussed in the context of the customer's application and the salesperson's understanding of the business user's industry. Generally, business users need to have the benefits of productivity, staying in touch for decisions, returning phone calls and messages, and conducting business over the phone personalized to their situation. Where personal use is involved, benefits of avoiding second trips on errands, security in the case of road emergencies, and time savings should be emphasized.

4. Closing: The sales presentation process can be described as a series of answering objections and test closes. If the prospect is responding to a specific offer, it can be a very brief presentation if they are ready to buy. Hesitant buyers can be assured that leasing and economy service plans can make their decision as tentative and non-commital as they desire.

5. Simplicity: The importance of the simplicity of the presentation (the need to answer questions honestly and to avoid dragging out the meeting once the sale has been consummated) cannot be over-emphasized. Optional service features, such as custom calling features and voice messaging, should only be sold if brought up in the context of the discussion.

RETAIL SALES

The retail sales operation is concerned almost exclusively with external retailers, such as automobile dealers, and has completely different problems from the mainstream direct sales operation. Little control can be exerted on such channels to change the way they do business. They use their own advertising, and the advertising and sales efforts are rarely directed specifically at cellular, but at a broad range of electronic, communications, or another larger product line.

Different Customers

The customers who approach such channels are also different. While some are business users looking for a deal on a cellular phone, more and more are uninitiated consumers. They do not approach the cellular decision from a business productivity angle and know far less about cellular before coming to the sales location. At the same time, the average retail salesperson's knowledge of cellular is far below that of the direct salesperson whose efforts are dedicated to cellular only.

Tactical Management

Where possible, a separate sales support manager who understands retail merchandising should be directed to help retailers manage cellular sales. The primary weakness in retail sales is the lack of information or conflicting information, which uninitiated sales prospects receive about such basic subjects as service pricing and coverage, or the advantages and disadvantages of mobile-installed units versus portables and transportables. When we consider the turnover of retail salespeople and the number of products they sell, the job of training may seem impossible. The value of making certain point-of-purchase literature and reference materials are available for salespeople, is paramount in this environment, as is on-the-spot training for salespeople on the floor. This is easily a full-time job in itself.

The 80/20 rule applies to retail stores and other non-traditional channels as much as anywhere else. Most sales will come from very few outlets. Success will not come from the facilities of one particular chain, but from a few highly motivated, well-supported locations of each company. This can be propagated to other locations only with a combination of the enthusiasm of the local manager and the support of the operation by the carrier.

CHAPTER 11

CELLULAR SALES PRODUCTIVITY[1]

Sales productivity is a major concern of all cellular players. While sales managers work hard to increase the number of sales per salesperson, there are other ways to analyze and improve sales productivity across all sales channels. One method is to analyze the productivity of each sales channel and optimize a mix of several sales channels based on their efficiency and appropriateness for specific kinds of customers.

Manufacturing businesses can associate current sales revenue with current sales and marketing costs, because most of the revenue from the sale of a product is immediately reflected in the income statement. This is more difficult to do with a subscriber service business like cellular, but it is necessary to

---

[1] Adapted from "Cellular Sales Productivity," in *Cellular Business*, April, 1992. Intertec Publishing Company, Overland Park, Kansas.

understand the relationship between sales costs and revenue in order to determine if its sales cost is an efficient generator of net income. We will use a cellular carrier here to illustrate how such an analysis is done, but resellers, agents, and other cellular players can use a similar method.

The best way to measure the efficiency of the sales effort of a subscriber service is to compare the current sales cost with the revenue that will result from the sale over the time that the average customer keeps it. This most closely relates the sales effort to the revenue it generates, in a way similar to the sale of a manufactured product. An average customer life of 30 months equates to an attrition rate of 3.33% of the customer base per month. If an average customer's bill is $80 per month, then the cost of each sale in the current month will yield 30×80 or $2,400 at retail in the future. We will call this the customer-life revenue, to distinguish it from current revenue on the income statement.

The cellular carrier must determine the cost of selling through each sales channel (direct, agent, dealer, reseller, etc.) and compare it to the profit that will result from the $2,400 in revenue it generates. A typical income statement for a cellular carrier might look like Figure 11–1, based on percent of revenue.

ALPHA CELLULAR COMPANY		
Income Statement as Percent of Revenue		
Revenue		100%
Expenses		60%
	Sales/marketing	13%
	Administration	7%
	Operations	20%
	Depreciation	20%
Gross operating income		40%

Figure 11–1   *A carrier income statement*

CURRENT VERSUS FUTURE REVENUE

While it is important to keep all categories of expense under control, the sales expense of this period does not generate this month's revenue, but future revenue. Minimizing sales/marketing expense as a percent of current revenue saves cost, but also reduces the ability to generate future revenue. The income statement in Figure 11–1 reflects that any revenue, current or future, that Alpha Cellular generates produces only 40% of revenue in operating income, before taking into account the discounting of future revenue streams. In our example, $2,400 of customer-life revenue will produce only 40%, or $960, in future operating income. If the carrier really wants to achieve an operating income of 40% of revenue within this cost structure, its long-run sales/marketing expenses must remain at or below 13% of revenue, as shown in the income statement. To accomplish this, the sales cost must also not exceed 13% of the customer-life revenue of $2,400, or $312. How many cellular carriers can say they are really doing this?

Unfortunately, many operators are so driven by subscriber numbers that they don't afford themselves this kind of rational analysis. The sales team is directed to get subscribers of any kind at any cost, and the financial manager and general manager wring their hands trying to cut costs indiscriminately.

Channel Income Statement

One of the first steps toward such a rational analysis is to attempt to construct an income statement for each sales channel. There should be records that show which sales channel has sold each customer and the current revenue for the set of customers attributed to each channel. Marketing costs such as advertising, which may be associated with multiple sales channels, may be allocated based on the proportional revenue or direct cost of each channel to which it applies. It is a real eye-opener to see the revenue and cost of each channel broken out separately.

The easiest division of channels is to separate the reseller channel from the aggregate retail channels (direct, agent, dealer). While the reseller channel may produce lower revenues at a lower wholesale price, its costs should be lower also. It is possible that the reseller channel may be more profitable than the retail channel. Not only are retail sales commissions saved when sales are made through the reseller channel, but the customer support function, billing function, and collection/bad debt costs are all lower. For comparison purposes, only those costs which differ among channels need be represented. Thus, operations, administration, maintenance, and depreciation expense per customer, for example, might be assumed to be the same in all channels. An illustrative version of such a statement is shown in Figure 11–2.

CALCULATIONS

The total channel cost can be divided by the gross sales gain to determine the cost per new subscriber by channel. However, not all customers are of equal worth. The total customer base for each channel should be divided into its revenue stream to derive current revenue/customer by channel to see what quality of customer each channel is producing in terms of revenue per month. This is then multiplied by the average customer life to determine the customer-life revenue. The customer-life revenue, remember, is not the same as the income statement revenue shown in Figure 11–2. It is merely a measure of channel efficiency when compared with the cost of the channel per sale.

The reseller average revenue per customer is $128,000 ÷ 2,000 or $64/month. The customer-life revenue is 64 × 30 or $1,920. For the retail channel, the average revenue per customer is $80 × 30 or $2,400.

CALCULATIONS

ALPHA CELLULAR COMPANY			
Monthly Income Statement by Channel			
Reseller Channel		**Sales Summary**	
Revenues		$128,000	Gross gain	40
Expenses		4,700	Customers	2,000
Marketing	$2,200		Rev/customer	$64
Customer service	1,200		Cost/gain	$118
Billing/collections	200		Customer-life revenue	$1,920
Sales	1,100		Rev/cost ratio	16
Revenue less channel expenses		$123,300		
Retail Channel		**Sales Summary**	
Revenues		$1,120,000	Gross gain	400
Expenses		224,000	Customers	14,000
Marketing	$57,000		Rev/customer	$80
Customer service	15,000		Cost/gain	$560
Billing	32,000		Customer-life revenue	$2,400
Sales	120,000		Rev/cost ratio	4
Revenue less channel expenses		$896,000		

Figure 11–2 *A montly income statement by channel*

ANALYSIS

In the illustrative monthly statements in Figure 11–2 for the composite of retail channels versus the wholesale or reseller channel, the retail channel produces far more revenue and customers than does the reseller channel. However, the cost per new subscriber is much higher than in the reseller channel. If these results were not separated by channel, the combined results would show a gross operating income of about 38% of revenue, close to our composite example. In reality, however, the retail channel is operating at a level of only about 33% profit, while the reseller channel is at 50%.

Revenue/Cost Ratio

The most important number is the revenue/cost ratio. The ratio of customer-life revenue to sales cost for the reseller channel is 16, while it is only 4 for the retail channel. The reseller channel is much more efficient than the retail channel.

How would you feel, as general manager, if you knew that it took $560 in sales costs to generate a gross income over 30 months of only $960? You may find that out, if, as the manager of retail channels, your results are like the retail illustration shown here. Instead of gaining customers at any cost, you might determine that unless low-usage consumers were willing to buy directly over the phone in response to an advertisement, you may not want to attract them with $200 cellular units. You may consider structuring commissions and bonuses based on the market segment of the customer.

This invites the same type of analysis of each individual retail channel—direct, agents, dealers, and retailers—to the degree that the information is reasonably easy to obtain. This would show which of the retail channels is the high-cost culprit, or if they all are. Those channels producing the highest-revenue customers would also be revealed. Channels producing customers at good revenue/cost ratios should be expanded, but only if market research shows that there is significant potential left in that market segment. It must also be considered that a channel cannot be expanded if there is little talent available with which to expand it. Channels that are high in cost and are not concentrating on high-revenue customers will be revealed by a low ratio. Expensive channels may not necessarily be the least efficient if they are producing high-revenue customers.

Expensive—But Worth the Cost?

The direct sales force is probably the most expensive sales channel. It should be matched to those corporate customers who need the kind of support they can give, and whose usage characteristics can support those costs with high revenue. Since this channel is normally associated with premium corporate accounts, it is often given the most marketing attention, the highest quality sales leads, the most professional sales resources, and other expensive support, which is difficult to assess objectively without careful analysis. However, this channel may pull in customers who not only may have higher monthly usage, but also may be more loyal and stay on the service longer to produce more revenue.

If possible, a different customer life should be calculated for each channel and used in the customer-life revenue calculation. This would show the truest picture of the quality of each channel's customers. A newer channel may produce distorted results in longer customer lives, because little customer attrition has set in yet. If a channel is selling to a special market segment, it may be a characteristic of the segment which causes greater loyalty, rather than a reflection on the quality of the sales effort of the channel currently selling to it. So it is also important to measure revenue and customer life by market segment, regardless of the channel selling to it.

Agents and dealers often bring in the bulk of new customers in the largest market segments. Commission levels may make the channel very expensive in a competitive environment. The opportunity to bring in retail storefront channels (or other channel alternatives) at lower commission levels for low-usage consumer segment customers is an option, and may be an attractive channel to these consumers. But it must be supported by a marketing program that matches the needs and costs of this segment and channel.

Matching Segments to Channels

As cellular grows, it must continue to attract more users from diverse market segments. In the larger view, the current business environment is experiencing a fragmentation of traditional market segments into smaller and smaller subunits—even "niches." Each of these subsegments must be attracted to cellular according to its needs and applications, but also with attention to its channel preference, and at a channel cost matched to its contribution to the business. Sales and marketing managers must become more attuned to the financial realities of marketing, rather than trying to attract any type of cellular customer at any cost.

As cellular subscriber numbers expand, more and more marginal users are attracted, which generates less than average revenues per subscriber. Therefore, the carriers must begin to understand which market segments generate the most revenue, and direct their highest-cost sales channels to those potential users. Other segments that generate less revenue must be approached through lower-cost sales channels.

If we understand the revenue per customer for each market segment, the cost per sale of each channel, and the channel preference of each segment, we can put together a matrix that matches channels to segments and exposes gaps in our approach, such as the one for Alpha Cellular (Figure 11–3). Only a few significant segments are shown in the figure for illustration. The largest would probably be the small business segment, which has not been included. The monthly revenue is multiplied by 30 months and divided by the cost per sale to calculate the customer-life revenue/sales cost ratio, referred to in the figure as "Ratio."

Indications for Action

This illustrative analysis for Alpha Cellular shows that the direct sales force costs too much to support the current corporate (or large business) segment, because the revenue is not high enough to support the cost. Some of this business should be switched to agents unless another alternative is proposed. The consumer market has an even lower revenue per customer, and desperately needs a lower-cost retailer-type channel, which research might show the customer would probably prefer. The reseller channel cost is so low that it should be expanded to attract other segments that are not attractive to retail channels or hard to reach. It may be a candidate channel for the construction segment, whose revenue does not support the sales cost in this market. A radio dealer who specializes in construction could become a successful cellular reseller with its existing customer base.

These are just a few of the actions that might be suggested by the analysis for a particular operation. Other cellular players would analyze a similar breakdown of their own channels and segments. An agent, for example, might want to do the same analysis to compare its direct sales, telemarketing, and dealer support channels.

Benefits

The process of producing and analyzing this data is as valuable as the result. Just realizing the need to see how sales channel effectiveness, sales channel cost, and segment revenue interact will produce large benefits in changes that increase both profitability and sales levels. Such changes must carefully consider customer needs and other factors, like the ability of a channel to expand.

Traditionally, marketing and sales have been asked to meet revenue and customer gain goals under budget control. They are being asked more often to directly relate their efforts to productivity and profit. This becomes more necessary as markets fragment and margins shrink. Such an approach will not only improve profitability, but also improve the quality of the customer relationship and the effectiveness of the sales channels, while introducing better communications between the marketing and financial people.

CHANNEL ANALYSIS					
ALPHA CELLULAR COMPANY					
MARKET SEGMENT					
BUSINESS					
	Corporate	Prof. Services	Real Estate	Constr.	Consumer
Monthly rev/cust	$200	$385	$270	$140	$60
Current channel	Direct	Agent	Package reseller	Two-way dealer	Agent
Current base	2,000	400	325	210	460
Potential	35,000	3,500	1,200	625	140,000
Cost/sale	$740	$560	$110	$450	$560
Ratio	8	21	74	9	3
Gap	High cost			Low rev	High cost/ low rev
Preferred Channel	Direct	Agent	Reseller	?	Retailer
Recommended channel	Agent	O.K.	Expand	Drop	Low comm. retailer

Figure 11–3   *A channel/segment matrix*

CHAPTER 12

❖ THE LIMITS TO GROWTH[1]

Sales managers are under constant pressure to achieve subscriber gains. It is unfortunate that the sales function is given the responsibility for net gain, since sales has little control over most customer losses. While sales can do more to make new sales "stick," and to keep existing customers satisfied, much of the problem has to do with uncontrollable churn. But this is not about churn—it's about keeping growth on track.

While sales managers are used to having their heads banged together for better sales results, their bosses don't always understand that a large part of the problem in growing the customer base has nothing at all to do with sales. Sales managers may want to tear out this chapter and show it to their general managers, if only to vent their frustration.

---

[1] Adapted from "Limits to Growth," in *Myths, Tips, & Facts*, a supplement to *Cellular Business*, July, 1992. Intertec Publishing Company, Overland Park, Kansas.

THE SYMPTOMS

When a cellular market first starts up, almost none of the customers leave the service until about 18 months after service begins. Churn isn't even noticed until about three years have elapsed, as a monthly attrition rate of 2-4% becomes a substantial number when applied to a customer base of reasonable size.

If we do not recognize this as a phenomenon from the beginning, it may appear that the sales productivity is slowing down, when in fact it is improving. So we need to break down net gain into its component parts: first, the gross gains in customers due to the sales effort, and second, the natural loss of older customers from the base due to customers changing jobs, moving, or becoming dissatisfied and moving to another carrier.

THE PROBLEM

If the customer base continues to grow but the sales effort stays constant, net gain eventually will be sharply reduced through no fault of the sales effort. This is the basic premise of what I call the "Steuernagel Syndrome":

> *Given any fixed amount of sales effort and any churn rate, net customer gain will virtually reach zero in a few years.*

A simple model that anyone can create in a few minutes with a spreadsheet will demonstrate this phenomenon. Just enter a constant sales rate of, say, 100 gross adds per month. Each month, add these to a total; then apply an attrition rate to the total (e.g., 3.33%) and subtract it from the total as churn; then repeat the process over many months.

Using these numbers, Figure 12–1 shows what happens.

MONTH	ADDS	LOSSES	IN-SERVICE	NET GAIN	GROWTH RATE
1	100	0	100	100	100%
2	100	3	197	97	97%
3	100	6	291	94	48%
.					
.					
.					
12	100	30	1,009	70	7%
24	100	54	1,684	46	3%
35	100	68	2,104	32	2%
36	100	69	2,135	31	1%

Figure 12-1   *At any fixed sales rate, customer growth will eventually reach 0.*

From the figure, we can see that after only 12 months, the net gain is only 70, or about 2/3 of the sales rate of 100. After 24 months, we realize a net gain of only 46% of the adds; and after 36 months, a fixed sales capacity of 100/month produces a net gain of only 29 customers and a growth rate of just 1%. While the sales force can contribute to the effort to lower churn through customer support, most of the losses are probably not due to customer dissatisfaction.

THE SOLUTION

Once we realize that we can only count on a customer for a certain term (a 3.33% monthly churn rate is equivalent to a 30-month customer life), we can recognize this in our financial and sales planning, and address the problem by applying the corollary to the Steuernagel Syndrome:

Sales capacity must continually expand to achieve growth in the customer base.

We can calculate how rapidly the sales rate needs to climb in order to maintain a constant growth rate, but we must also consider the following parameters in expanding sales capacity:

1. Sales productivity, or the number of sales per sales cost dollar expended, must remain constant or increase.
2. Existing sales channels cannot be expanded indefinitely; new sales channels must be found and exploited.
3. Target market segments and geographic territories must be broken down more narrowly among the sales channels, and new target markets uncovered.
4. New applications, service features, pricing plans, and sales opportunities must be developed to match new channels and customer segments.

GROWING THE SALES CAPACITY

Month 36 of service in our example was the first month in which the growth rate of the customer base dipped to 1%. In order to keep the growth rate at, say, 2% per month, or about 24% per year, we merely "size" the sales capacity required, and grow it accordingly, instead of merely keeping the sales capacity constant. Using our example again, instead of maintaining the sales capacity or rate at 100 per month, starting in month 36 we calculate the sales rate required to be equal to the attrition, or churn rate, plus the desired growth rate, times the customer base for month 35.

(Growth Rate + Churn Rate) × Customer Base
= Sales Capacity

(2% + 3.3%) × 2,104 = 112

So in month 36 we must increase our sales capacity from 100 to 112 and keep our customer base growth rate at 2%. Once we make this adjustment, all we have to do is keep growing the sales force at the same rate as the desired growth rate—but we must constantly increase the sales capacity thereafter, or the growth rate will drop to zero. In our example, this means that we will have to more than double our sales force every three years.

OTHER USES

Some of the earliest cellular markets arrived at these conclusions through trial and error. As the sales capacity is expanded, the existing salespeople, agents, and resellers may express concern that the expansion of the sales force is unnecessary and competes with their efforts. The need to diversify the sales effort and grow it at a management-controlled rate, requires the carrier to add sales channels and channel members and still grow existing ones. The demonstration shown here can help allay the fears of in-place sales channels, through providing an understanding of the phenomenon.

"Topping Out"

If we look at an agent instead of a carrier, we can see the same flattening of growth through this analysis. If a new, conventional cellular agent starts up with a sales force of 20 who each produce 25 sales per month, its nominal sales rate is about 500 per month. Yet this analysis will show that unless productivity increases, the sales force is enlarged, or the agent adds other sales methods, that the agent will "top out" in three or four years at 10,000-12,000 customers.

The Limits to Growth

The flattening of the growth rate for an agent does not mean that performance is at fault. A new agent might look attractive compared to an existing agent because churn has not yet become a factor, but in reality is not performing any better than the in-place channel. The solution is to include this analysis in the forecasting of gains over the next 3-5 years, and to constantly seek additional channel members and new channels that meet the needs of emerging market segments, while supporting existing channels in productivity improvements and new sales methods.

This analysis can also be used as the basis of an analysis to understand the reasons for rapidly declining customer base growth rates as a cellular system matures. It also serves as a sensitivity analysis for the effects of increased and decreased churn and to demonstrate what portion of increased sales capacity can be met through productivity increases.

Most intriguing are the implications for the sales force size if a carrier in a top 10 market expands its nominal service capacity through digital technology by a factor of 10 or more. Our model may infer that finding sales resources for filling a digital cellular system may be even more difficult than deciding on the technology standard has been, and that it may be unfeasible, from a sales perspective, to fill a vastly expanded capacity in a reasonable time frame.

CHAPTER 13

SELLING THE CONVERSION TO DIGITAL

THE NEED FOR DIGITAL: CAPACITY

One of the largest problems looming ominously directly in front of major carriers is the conversion of users from analog to digital. In the near term, this problem is most imminent for carriers in the largest cities, where success in gaining subscribers has choked conventional analog systems, forcing them to move as early as possible to narrowband and digital solutions which increase capacity.

This has in turn ignited both objective and self-serving arguments about the benefits of TDMA (Time Division Multiple Access) versus CDMA (Code Division Multiple Access) and even a few other digital technologies. Most of the these arguments deal with trade-offs of quality versus capacity, or limitations of technologies in commercial environments. It is important that such discussions result in the clarification of these tradeoffs to reach eventual agreement on technical standards. The concentration on the solution of the technical problems has made it the crux of the marketing problem also.

As carriers consider the phasing-in of digital under various technologies and rates of change, important questions about the way the system will behave are addressed. The most important considerations, those of ensuring that customers using digital or analog service will experience no change in service availability or quality as cells and systems change dynamically over many months, must be worked out.

Making Customers Part of the Problem

These technical problems of standards choices, system design, and operations and engineering problems of the changeover have unfortunately occupied carriers so much that they can end up with only one thought regarding the marketing of the new technology:

> *How are we going to convince customers to buy expensive new dual-mode phones and migrate to digital to make use of all this new capacity we've created?*

The fact that the customers will have to buy new cellular phones in order to allow the carrier to make use of the new technology does not mean that they have to consider the technology changeover to be a capacity problem. There is no need to go to customers with hat in hand and say, "Will you please buy a new cellular phone if I subsidize $400 of the cost so I can start increasing capacity?" There is no reason why customers have to even consider the carrier's problem. They didn't create it.

In addition, it is difficult to work out the economics of paying $400 to get existing customers to use digital. You get no additional revenue, and you don't displace any costs.

It is hard to understand the logic in going after heavy users first, and inconveniencing these, your best customers, based on the argument that then you only have to convert half as many to use the same digital capacity as light users (whom you don't value as much and yet are willing to leave out of your problem). Do your best customers deserve such treatment?

But already carriers have committed to tens of thousands of dual-mode analog/digital telephones, assuming that they will have to subsidize them, and still beg customers to help them with their problem for the greater good.

A SOLUTION

A better way to think about the changeover is to remove the shackles of the capacity problem. Don't assume you have to subsidize the phone and beg customers to take them. Turn the problem into a marketing opportunity.

First, consider that if you do not target existing users, but concentrate on new customers for digital, that you avoid the cost of converting existing users (without any increase in customers or usage revenue). Second, for the purposes of marketing, forget the capacity problem and concentrate on benefits of the digital service for users, such as clarity, faster access, fewer dropped calls, security, and advanced intelligent features. Third, understand the marketing value of merely telling new users that your cellular system now has "end-to-end digital technology," "CD-quality sound," or "air-tight security." Fourth, have one of your favorite cellular telephone manufacturers design a cosmetically distinctive dual-mode phone with some genuinely new features, and have them create a line of models, complete with literature and subordinate logo for the phone that says, for example, *"DIGITAL READY"* or *"ALL DIG-ITAL."*

Fifth, forget about your ego urging you to promote your system, and promote the telephone instead as "the next generation of cellular," whose advanced technology can only be used to advantage on your system (if you are first in the marketplace; if you are second, the other carrier will already have spearheaded digital for you in this manner). Sixth, offer the phone at a slight premium in price over similar conventional analog phones, to highlight its increased value. Seventh, create a price/premium promotion in connection with the new telephone. Offer it through both your internal and your external channel partners, but with a limited-time price offer (above its analog equivalent), contest, or a free premium.

The best solution is to forget about the technological changeover problem that originated the conversion, and think instead about the marvelous customer benefits and marketing opportunity it creates. The above are just a few key elements of a marketing program that not only provides an excellent premise for a promotion to increase sales to new customers in its own right, but it just happens to solve the digital conversion problem.

The worst solution, which many carriers appear determined to pursue, is to inconvenience their best subscribers, bribe unwilling users to change telephones, spend huge amounts of money with no revenue gain, and treat the conversion as a problem to be shared with customers.

CHAPTER 14

SWORDS INTO PLOWSHARES: Cooperation and the Evolution of PCS

A NEW APPROACH

The success of cellular should be credited to the participation of multiple players and the diversity of marketing competition, even more than the market need. Instead of figuring out ways to reduce competition for spectrum and sales, each of the potential players in PCS should recognize that the market penetration of cellular through multiple players is 4-5 times the original forecasts on the order of 1.5% that each cellular carrier hoped to garner by excluding others from competition.

The crux of success in cellular has been the proliferation of competition in cellular telephones and subscriber sales, in spite of efforts to contain it. In order for PCS to succeed, the players should embrace competition in step with the regulatory initiatives which favor competition.

How does working together alternately as partners and competitors achieve a whole which is much greater than the sum of its parts? Here are a few examples for cellular and PCS.

1. PCS can coexist with cellular in locations where there is significant non-mobile need for wireless service. PCS service supplements cellular service in congested areas where further cellular capacity increases would entail great expense or unfeasibly small cells, and complements cellular service in closed buildings and fixed applications.

2. PCS can complement cellular where the provision for peak traffic compared with average traffic would require a huge capital expense for cellular, yet be economic for a PCS microcell.

3. Dual cellular/PCS portable telephones can enable existing cellular users to opt for lower-priced service and off-load cellular cells where there is congestion. The cellular carrier can purchase PCS service for this purpose at wholesale from the PCS carrier, where it is not a PCS carrier itself. The PCS carrier can become a cellular reseller and also integrate the services into a full product line.

4. PCS, at lower rates and lower functionality than cellular, improves the market positioning of cellular as a premium service for those who need it.

5. PCS provides a product line extension to cellular sales entities (whether they are part of the internal sales force, agents, dealers, or resellers) plus an additional choice to prospective buyers. An important classic sales ploy is to take a potential customer who isn't sure he wants a car and turn it into a choice between two different models.

6. A dual digital cellular/PCS portable telephone provides a mechanism for converting certain cellular customers to new cellular protocols as a marketing opportunity, rather than as a burdensome technology changeover problem.

7. PCS is a market entry vehicle for cellular. Newcomers to wireless can first subscribe to PCS at reduced prices, using it not as "poor-man's cellular," but as a premium cordless and public telephone. As they become accus-

tomed to its value, they will become more communications-intensive and become good customers for cellular over a wider area.

Some of these options presuppose a partnership among cellular and PCS providers. In the broadest sense, there would be an actual integration of the network and billing operations of cellular and PCS providers, giving users automatic or chosen access preferences, options for disposition of unanswered incoming cellular calls to PCS and vice versa, and forwarding to paging or voice messaging, answering service, or an office. An integrated bill could also be provided.

The players are taking sides as Personal Communications Service (PCS) unfolds, even before a full understanding of the service offering and licensing procedure is available. Let's take a look at the real and imagined perceptions of some of the players and see how these can be rearranged into a PCS industry that capitalizes on the strengths of the players and how they may complement each other.

CELLULAR

Cellular carriers uniformly voice the opinion that cellular can accommodate all of the needs for any market segment or application of PCS. Because many are about to invest millions in digital technology which will increase their capacity by three to ten times, they see PCS as unwelcome competition for the attention of their potential customers as they try to fill this capacity with customer usage.

At the same time, they see a big problem in convincing customers to buy digital—capable TDMA/CDMA cellular phones at increased prices to make use of this additional capacity, while PCS units promise to be priced even lower than analog cellular. It's not surprising that they feel threatened.

The cellular industry may feel that their arguments are more credible if they refuse to acknowledge the inevitability of the announcements by the FCC, of the allocation of spectrum and of rules for accepting license applications, before they start talking about their role in non-cellular PCS in earnest, especially if they are allowed to apply for licenses. The flood of applications for PCS licenses from cellular carriers will belie their current arguments. It is ironic that an industry can turn from an offensive, entrepreneurial fight for a place in the sun to a defensive, entrenched industry position in a few short years.

LOCAL EXCHANGE COMPANIES (LECS)

Telephone companies have been very slow to incorporate wireless local exchange access alternatives, even in remote areas where the economies are clear. There are perhaps two reasons for this reluctance. First, their engineering tradition and practices are steeped in wire technology, and there are certainly issues of reliability, feature transparency, and network integration to resolve. On the other hand, telephone companies perhaps concentrate too much on wireless's cost parity with wire, rather than the marketing of premium access services.

As long as they view wireless as a special extension of the embedded network, they will not have the same open perspective toward PCS that a new entrant has. Of course, they seem open to the advent of PCS as a separate market like cellular, and are planning appropriately for access products to accommodate PCS service providers' needs. As local telephone services open to competition, they are recognizing the possibility that PCS, cellular, operator services, alternative carrier services, and other players are essentially telecommunications traffic aggregators who have other carriers to choose from.

As a PCS carrier, local telephone companies can offer premium residential cordless service as a way to move upscale residential telephone subscribers from subsidized basic wire access at a loss, to premium access service at a profit.

As a reseller, a telephone company can offer basic service to remote neighborhoods through PCS, lowering their subsidy cost. A relocatable box at the entry point to the residence makes the service transparent to the subscriber, who sees a normal basic access line for existing CPE. If the LEC can buy access for $10/month from the PCS provider and the latter uses the LEC as primary carrier, the telephone company gets all of its traffic while reducing its access costs. While there is the additional cost of the box at the entry point, the telephone company avoids competitive loss of traffic to intra-LATA competition.

This arrangement eliminates the telephone company's problem of installing non-reusable wire facilities to each subscriber, while it deters traffic competition and aggregates usage for the network. The PCS provider is guaranteed a subscriber without marketing and sales costs, and can still provide premium cordless service to some subscribers in the neighborhood in a joint sales arrangement with the telephone company. If the customer opts for premium cordless service, the box is removed from the entry point and reused, and the subscriber buys a compatible cordless portable that addresses the neighborhood base station directly, without changing telephone numbers. This customer thus would pay a higher price for the value of premium cordless service, while the PCS provider and the LEC lower their costs.

EQUIPMENT PROVIDERS

One of the basic truths of cellular is that the cellular telephone, not the cellular carrier, has been the focus of the customer decision process since its inception. Much to the consternation of cellular carriers and resellers, the prospective cellular customer finds the best deal on the telephone of his or her choice, and usually subscribes to whatever provider is associated with the seller, rarely considering the advantages of one carrier or reseller over another.

As with cellular, the most successful PCS telephones will be the ones that have the most wanted features or the lowest price; there is no middle ground. The manufacturer needs access to customers through the sales channels of the PCS carrier, and the PCS carrier needs a manufacturer who will cooperate in promotions centered on the telephone but designed to attract service customers. While features and price play large roles, the ability of the manufacturer to respond to the carrier's marketing plan are essential: advertising co-op, sales support, financing, warranty service, and a continuous stream of promotable new models and features.

On the surface, vendors of network equipment and telephones stand to gain the most from the advent of PCS, because of the diversity of the kinds of equipment required. Competing technologies require analog, TDMA, or CDMA capabilities, plus a range of single- or dual-mode telephones. However, competition in these areas in the cellular arena have made profits elusive for manufacturers, and providers are much more cautious about creative financing for network equipment and low prices for telephones in anticipation of volume than they were with cellular.

CABLE TELEVISION INTERESTS

Cable television interests, like LECs, appear to approach PCS initially as an attempt to leverage their investment in cable plant, rather than as a marketing opportunity. While PCS as an alternative access in residential neighborhoods is only one application of the technology, it is seen as an extension of their existing plant into telecommunications applications, rather than a pure opportunity based on an objective evaluation of where market need is highest.

To the extent that the LEC does not opt for one of the above arrangements, cable interests can provide such access in competition with the LEC, as a PCS carrier or as a partner with one. An IEC (Inter-Exchange Carrier) might also participate in such bypass of local access. The cable carrier might alternatively be the LEC's partner and provide facilities from its "head end" to the neighborhood PCS base station.

PAGING CARRIERS

The paging user appears to be the closest natural target market segment for PCS. The embedded base of customers of paging carriers and their sales channels may be the best way for PCS carriers to address potential customers economically, and add a product line and important adjunct service to the paging carrier's offerings. A mutual alliance or sales agreement between PCS and paging providers is a virtual necessity for a public telepoint network.

NEW PCS CARRIER STARTUPS

The extraordinary success of cellular provides a historical context that urges a broad spectrum of entrepreneurs and existing companies to invest in PCS. Although they present all kinds of arguments promoting PCS, their collective attitude is:

1. that the ability to compete for spectrum as a scarce resource is a gold mine, no matter what service it is initially connected with, and
2. that cellular carriers and wireline telephone companies should be excluded from license competition, an attitude at the opposite extreme from cellular carriers.

PCS technology and the economics of the public telepoint applications of PCS may preclude ubiquitous, market-wide PCS service. However, as long as the telephone unit works on any PCS carrier's system through a Common Air Interface (CAI) standard, each of the three to five PCS carriers in any one market can pursue complementary deployment strategies, applications, and market segments to make the service and each of its carriers a success. If all carriers complement each other's locations, including some overlap, the public will end up with adequate service at the right number of locations. In addition, if one carrier emphasizes in-building wireless service, another neighborhood residential service, and a third attracts commercial applications (shopping malls, commercial strips, transportation centers, etc.), the overlapping requirements of users will be served.

It is hard to imagine, however, that this scenario might come about. More probably, each carrier will attempt to control an entire city by itself without acknowledging the others, and design a system to cover every nook and cranny. They will strive to be entirely self-sufficient in marketing and issue low-ball forecasts to make the industry look unattractive to other potential players.

CONCLUSION

As evidenced by the facts that cellular carriers who attempt to control their market themselves generally do poorly and those who nurture a diverse yet competitive environment of multiple players have excelled, the secret in PCS will be to cooperate with the multiple dimensions of competition in the markets rather than deny it. Unlike cellular, however, not only will marketing partnerships be necessary, but PCS carriers and multiple transmission technologies will have to learn to complement each other rather than compete head-on for the same customer (Figure 14-1).

Some regulatory help is also required in the short term to allow full participation of all of the players in case PCS licensees do not evidence such an open approach. First, all players must be allowed to compete for licenses, including cellular, telephone, and new entrants. Second, a resale opportunity at wholesale prices should be required, which would encourage entry of a second tier of retail sales of the service and avoid a duopoly or oligopoly in each market. Third, help in rapidly establishing a standard, full cross-licensing of technologies, and without proprietary grandstanding, will accelerate the deployment of PCS in response to public need.

CONCLUSION

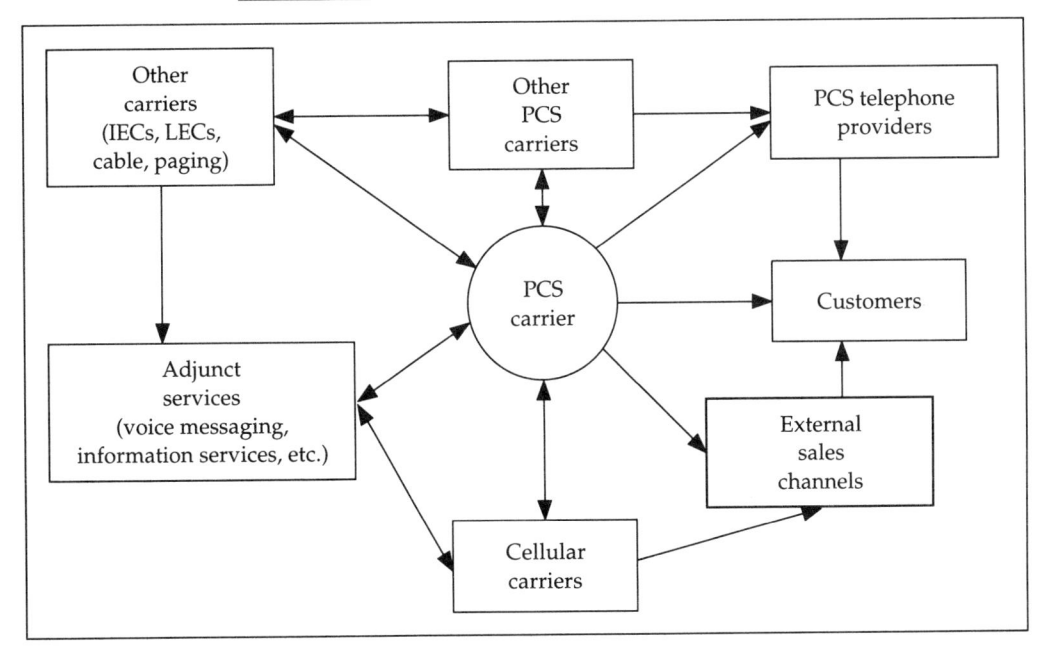

Figure 14-1 *Potential PCS relationships*

While this analysis of the perspectives of stakeholders in PCS may be incomplete and over-simplistic, it is valuable as a contrast to an environment where players can find a place to compete in PCS while complementing the positions of other players, rather than stepping on them.

INDEX

Adjunct services, 20
Advertising, 106. *See also* Promotion
  agent, 87
  carrier, 86
  cost allocation of, 151
  definition of, 103
  educational, 10, 106–7
  expense of, versus lead generation, 117
  measurement of, 108–9, 112–13
  media for, 110–11
  multiple, 107
  pitfalls in, 112–13
  promotional, 108
  radio, 111
  relative value of, 142
  single-minded proposition in, 109–10
  time-limited call to action in, 10, 109
Advertising-pull, 28, 75
Agents, 67–68, 80, 89
  advertising by, 87
  contracts with, 92–98
  financial statements of, 97–101
  income statement of, 98, 99, 100
  market share by, 78
  measurements used by, 119
  organization of, 86
  versus resellers, 85
  sales and marketing activities of, 86–88
  sales potential of, 155
Aggregators, 69
Airtime, 19–20
Alpha Cellular, 132–33, 151, 156
  sales analysis of, 157–58
Analog, conversion to digital, 165–68
Annual plan, 33, 44–48, 138
AT&T, 4
Auto-market research, 113–14
Awarding of licenses, 4–5

Basic pricing plan, 54, 56, 57, 58, 60, 65
Bell companies, 135
  restrictions on, 7
Billing
  cost of clearing roamer, to home carrier, 62
  custom services for, 21
Boutique versus industry, 130–31
Branding in competitive differentiation, 129–30
Broker-agents, 69–70, 82–83. *See also* Agents
  contract for, 92–93
Burning eyes syndrome, 112, 133
Business ethics, 137

Cable television interests, 174
California Public Utilities Commission (CPUC), 89
Callbacks, value of, 142
Call forwarding, 20, 60–61
Campaigns, 143
Carrier advertising, 86
Catalog, 123
CDMA (Code Division Multiple Access), 165, 174
Cellular buyer, profile of, 13
Cellular buying process, stages of, 10, 11
Cellular carriers, as personal communications service carrier, 171–72
Cellular customer, revenue per, 12–13
Cellular customer base, 1, 2
  growth of, 160–61
Cellular data transmission, 21–22
Cellular industry
  care of, 6
  growth in, 1–3, 159–64
  limits to, 164
Cellular licenses, awarding of, 4–5
Cellular market, 1–3
  changes in, 12–13
Cellular marketing
  costs in, 5
  framework, 6
  history of, 3–5
  regulatory constraints, 4–5
Cellular phones
  buyers versus users, 13–14
  customer benefits, 9–10
  evolution of, 9
  promoting, 104–6
  types of, 15–18
Cellular sales productivity, 149–50
  analysis, 154–58
  calculations, 152–53
  current versus future revenue, 151–52
Cellular service
  features of, 19–20
  product dimension of, 20
Cellular subscriber base, growth of, 1, 2
Cellular Telecommunications Industry Association (CTIA), 1

Cellular users, characteristics of, 13–14
Centralization, 32–34
Channels
  calculating total costs, 152
  conflict in, 81, 96
  diversification in, 26–27
  income statements in, 151–52
  matching segments to, 156
  mix, 77–78
  relative performance in, 119
Churn, 45, 98, 117–18, 160, 164
Circuit City, 72
Cold-calling, 81, 143
Common Air Interface (CAI) standard, 175
Company stores, 76
Competition
  disparagement of, 136
  identifying, 134–35
  on your own terms, 133–34
Competitive backlash, 135–36
Competitive behavior and analysis, 127, 131
  boutique versus industry, 130–31
  competitive differentiation, 128–30
  unique customer proposition, 132–33
Competitive differentiation, 128–30
Competitive information, overvaluing, 137
Competitor, identifying, 134–35
Contracts
  agent, 92–98
  broker-agent, 92–93
Coordination, 142
Corporate mission and objectives, 37–38
Corporate pricing plan, 54, 56, 57
CPE (Customer Premises Equipment), 4
CPE (Customer Provided Equipment), 7
Cream-skimming, 40
Crossover point, 58–60
Custom billing services, provision of, to multi-user corporate customers, 21
Custom calling features, 20
Customer benefits, 9–10
Customer gain
  versus marketing expense, 116–17
  versus sales expense, 116

INDEX

Customer-life revenue, 152
Customers, 88–89, 120–21
   measuring gain and usage levels, 121
   in retail sales, 146
Customer satisfaction, 117
   studies on, 123–25

Database marketing, 122, 123
Data service, 20
Dealers, 68–69
   market share by, 78
   sales potential of, 155
Decentralization, 5, 31–32
Digital, 3
   selling conversion to, 165–68
Diminishing returns, law of, 109
Direct dealers, 68
Direct mail, 75, 108, 123
Direct marketing, 74–75, 122, 123
Direct sales, 144
   market share by, 78
   sales planning in, 144
   sales presentation in, 145–46
Direct sales force, 25–26, 41, 78, 96
   costs of, 155
   need for, 27
Discount pricing, 43
Disparagement, 136
Distribution, 67. *See also* Sales channels
   in competitive differentiation, 129

Economy, 60
Economy pricing plan, 54–55, 56, 57
Educational advertising, 10, 106–7
Effectiveness, 119–20
Efficiency, 119–20
   measuring sales effort, 150
80/20 rule, 147
Equal access rules, 135
Equipment providers, 173–74
Equipment sourcing, 89
Exclusive dealer, 80
Exclusive sales agents, 41
Executive pricing plan, 54, 56, 57

FAX, 20
Federal Communications Commission (FCC)
   awarding of licenses, 4*n*
   requirement for resale, 70
Fixed cellular service, 20
Flattery, 136
Focus groups, 114
Follow-up service, 91

General manager, 30, 31
Geographic market, 29–30
Goals, distinguishing between, and strategies, 38
Gross customer gain, 118
Growth, defining goals in, 37
GTE, 84, 85

Headquarters marketing function, 33–34
Headstart provision, 5
Hybrid model, 77–78

IEC (Inter-Exchange Carrier), 174
Income statement
   of agent, 98, 99, 100
   in channel, 151–52
   pro forma, 48
Industry, versus boutique, 130–31
Information gathering, 122–23
Information services, 121, 128
Intangibles, promoting, 103–4
Intel, 104
Internal direct sales force/major accounts, 73–74

Key account, 73

Lead generation, 143
   versus advertising expense, 117
Leads, 118
   tracking, 141
Leased equipment, 89
Local Exchange Companies (LECS), as personal communications service carrier, 172–73

Local toll and message charges, 19
Longitudinal studies, 125

Mail-in replies, 109
Market-based pricing, 50
Marketing, 28–31
  costs of, 151
  increased role of, 27–28
  role of, in controlling revenue, 25
Marketing expense
  versus customer gain, 116–17
  versus revenue, 116
Marketing mix, 6
  and pricing, 51–52
Marketing objectives, 40
  strategies to achieve, 40–41
Marketing organization
  centralization, 32–34
  decentralization, 31–32
  evolution of sales, 25–28
  marketing function, 28–31
Marketing plan, annual, 33, 44–48
Marketing strategies, complementary, 43–44
Marketing success, general measures of, 120–21
Market management, 29
Market research, auto, 113–14
Market segmentation, 10–14, 22–24
Market share by channel, 78
Master agents, 69
Maytag Corporation, 129
McCaw, 5, 14, 85
Measurements
  of advertising, 108–9, 112–13
  basic rules of management, 118–19
  customer satisfaction, 123–25
  disciplines of, 115–19
  of marketing success, 120–21
  of promotion success, 122–23
  useful, 119–20
Media, 108, 110–11
Mission statements, 38–39
Mobile telephones, 3–4, 15–16
Monthly performance information, 96
Motorola, 16, 84, 85, 129

Multiple advertisements, 107
Multiple end-user providers, 130
Multiple pricing plans, 54–55
Multi-user corporate customers, provision of custom billing services to, 21

National cellular programs, 85
New customer gain, forecast for, 47–48
Newspaper, advertising in, 111
NonStop computing, 132
Non-wirelines (NWLs), 4–5, 135–36

Off-peak pricing, 62–63

PacTel, 5
Paging, 87–88, 175
Peak usage price, 63
Personal Communications Service (PCS), 3, 135, 171
  carrier startups, 175–76
  cellular carrier as, 171–72
  evolution of, 169–78
  local telephone companies as carrier, 172–73
  potential relationships, 177
  telephones in, 173–74
P&L (profit and loss) statement, 31, 32
Portable phones, 15, 16–18
Premium pricing plans, 20–21
Price elasticity, 63
Pricing, 49–50
  business assumptions for, 41–42
  comparative plan features, 55
  in competitive differentiation, 128–29
  crossover point, 58–60
  development of, 50–51
  elements in, 62–63
  and equipment, 89–90
  financial considerations, 62–66
  and marketing mix, 51–52
  multiple plans, 54–55
  pitfalls in, 55–57
  revenue relationships, 57
  roaming, 61–62
  and sales commissions, 60
  specific assumptions, 42–43

strategies in, 64–66
structure and levels, 52–60
vertical features and options, 60–61
wholesale, 57–58, 89
Productivity, 32
Product
    attributes of, 14
    definition of, 128
    management of, 28–29, 29
    versus services, 18–19
Pro forma income statement, 48
Promotion, 108. *See also* Advertising
    of cellular phones, 104–6
    in competitive differentiation, 128
    definition of, 103
    importance of tracking, 122–23
    of intangibles, 103–4
    measuring success of, 122–23
Public relations, 111–12
Pull marketing, 75

Radio advertisements, 111
Radio Shack, 72, 93
Rate plans, comparison of, 59
Referrals, 86
    relative value of, 142
Refund policy, 128
Resellers, 4, 70–71
    versus agents, 85
    average revenue per customer, 152
    market share by, 78
    and use of personal communications service, 172–73
    value of, 57
    wholesale sales by, 41–42, 83–84
Residuals, 82–83, 96
Retailers, 71–73, 89, 146
    customers in, 146
    market share by, 78
    tactical management in, 146–47
Revenue
    current versus future, 151–52
    per customer, 118
    growth of, 2–3, 120
    versus marketing expense, 116
    roamer, 61

role of marketing in controlling, 25
versus sales expense, 116
Revenue/cost ratio, 154–55
RFP (Request For Proposal), 14
Roaming, 61–62

Sales, evolution of, 25–28
Sales aggressiveness, strategy for, 138
Sales capacity, growth of, 162–63
Sales channels
    agency terms, 93–94
    agent contracts, 92–93
    agent financials, 97–101
    agents, 67–68, 78, 80, 86–88, 89, 92–98, 99, 100, 119, 155
    agents versus resellers, 85
    broker-agents, 69–70, 82–83
    channel conflict, 81
    channel mix, 77–78
    characteristics of, 78–80
    company stores, 76
    customers, 88–89
    dealers, 68–69, 78, 155
    direct marketing, 74–75, 122, 123
    internal direct sales force/major accounts, 73–74
    market share by channel, 78
    mix of, 41
    national cellular programs, 85
    pricing and equipment, 89–90
    resellers, 4, 41–42, 57, 70–71, 78, 83–84, 85, 152, 172–173
    retailers, 71–73, 78, 89, 146
    service, 90–91
    trends in, 91–92
Sales commissions, 81–83, 98
    in agent contract, 94–95
    costs of, 155–56
    payment of standard, 72
    and pricing plans, 60
    range of, 82, 94
    residuals in, 82–83
    structure of, 82
    trends in, 92
    up-front, 92, 96

Sales expense, 118–19
  versus customer gain, 116
  versus revenue, 116
Sales function in newer markets, 30–31
Sales managers, pressure on, to increase subscriber gain, 159
Sales planning, for direct sales, 144
Sales presentation, in direct sales, 145–46
Sales process
  campaigns in, 143
  coordination of, 142
  direct sales in, 144–46
  general considerations in, 141–43
  lead generation in, 143
  retail sales in, 146–47
  sales support in, 143
Sales-push, 28
Sales support, 143
Sears, 72, 93, 129
Security, 137
Segments, matching, to channels, 156
Service, 90–91
Services, versus products, 18–19
SMR (Specialized Mobile Radio), 135
Sprint, 128
Steuernagel Syndrome, 160, 162
Strategic objectives, 39
Strategic plan, 36–37
  development of, 36
  well-constructed, 138
Strategies, 39–40
  distinguishing between, and goals, 38
Subscriber services, marketing of, 18–19
System growth, 138–39

Tactical management, in retail sales, 146–47
Tandem, 132

Target market
  identifying, 10–12
  price structures for, 42–43
TDMA (Time Division Multiple Access), 165, 174
Telemarketing, 75–76, 123
  inbound, 75
  relative value of, 142
  outbound, 75
Telemetry service, 20
Television
  advertising on, 111
  cable interests in personal communication service, 174
Three-way calling, 60–61
Time-limited call to action, 10, 109
Topping out, 163
Transportable phones, 15, 16

Unique customer proposition, 132–33
Usage, 120–21

Value-added services, 87–88
Vertical features, 60–61, 128
Voice massaging, 20, 21, 121, 128
  pricing of, 56–57

Wholesale
  benefits of selling at, 83–84
  pricing, 57–58, 89
  profit in, 71
Willingness-to-pay (WTP), 50
Wireless data, 135
Wireless voice technologies, 3
Wirelines, 129
  restrictions on, 7
Wireline (WL) licensees, 4
Women, as potential market, 12